Made to Hear

MADE TO HEAR

Cochlear Implants and Raising Deaf Children

• • • • • • • • • •

LAURA MAULDIN

A Quadrant Book

University of Minnesota Press
Minneapolis • London

KH

Quadrant, a joint initiative of the University of Minnesota Press and the Institute for Advanced Study at the University of Minnesota, provides support for interdisciplinary scholarship within a new, more collaborative model of research and publication.

http://quadrant.umn.edu.

Sponsored by the Quadrant Health and Society group (advisory board: Susan Craddock, Jennifer Gunn, Alex Rothman, and Karen-Sue Taussig) and by the Center for Bioethics at the University of Minnesota.

Quadrant is generously funded by the Andrew W. Mellon Foundation.

Portions of chapters 1 and 2 were published in different forms in "Parents of Deaf Children with Cochlear Implants: A Study of Technology and Community," *Sociology of Health and Illness: A Journal of Medical Sociology* 34, no. 5 (2012): 1–15. A different version of chapter 4 was published as "Precarious Plasticity: Neuropolitics, Cochlear Implants, and the Redefinition of Deafness," *Science, Technology, and Human Values* 39, no. 1 (2014): 130–53. Portions of chapter 5 were published in "Cochlear Implants and the Mediated Classroom-Clinic: Communication Technologies and Co-operations across Multiple Industries," *Disability Studies Quarterly* 31, no. 4 (2011), http://dsq-sds.org/article/view/1713.

Published by the University of Minnesota Press
111 Third Avenue South, Suite 290
Minneapolis, MN 55401-2520
http://www.upress.umn.edu

Library of Congress Cataloging-in-Publication Data
Names: Mauldin, Laura, author.
Title: Made to hear : Cochlear implants and raising deaf children / Laura Mauldin.
Description: Minneapolis : Univ of Minnesota Press, 2016. | Series: A quadrant book | Includes bibliographical references and index.
Identifiers: LCCN 2015031932 | ISBN 978-0-8166-9724-3 (hc) | ISBN 978-0-8166-9725-0 (pb)
Subjects: LCSH: Hearing impaired children—Means of communication. | Deaf children—Family relationships. | Hearing disorders in children—Rehabilitation. Cochlear implants—Social aspects. | BISAC: SOCIAL SCIENCE / People with Disabilities. | SCIENCE / Philosophy & Social Aspects. | MEDICAL / Audiology & Speech Pathology.
Classification: LCC HV2391 .M38 2016 | DDC 649/.1512—dc23
LC record available at http://lccn.loc.gov/2015031932

Printed in the United States of America on acid-free paper

22 21 20 19 18 17 16 10 9 8 7 6 5 4 3 2 1

To the memory of my loving grandmother, Helen

Science is politics by other means.

• Sandra Harding, *Whose Science? Whose Knowledge? Thinking from Women's Lives*

For the ethnographer in a medical setting, health care workers are an exotic tribe. . . . At worst we are voyeurs seeking cheap thrills; at best we are witnesses reporting on the most profound dilemmas of the human condition.

• Charles Bosk, *The Fieldworker as Watcher and Witness*

CONTENTS

ABBREVIATIONS

AAA	American Academy of Audiology
AAP	American Association of Pediatrics
ABR	auditory brainstem response test
ADA	Americans with Disabilities Act
AGB	Alexander Graham Bell Association
ASHA	American Speech-Language-Hearing Association
ASL	American Sign Language
AVT	Auditory Verbal Therapy
CDC	Centers for Disease Control and Prevention
CI	cochlear implant
EHDI	Early Hearing Detection and Intervention
EI	Early Intervention
ENT	ear, nose, and throat
FDA	Food and Drug Administration
IFSP	Individualized Family Service Plan
JCIH	Joint Committee on Infant Hearing
LSL	listening and spoken language
NBHS	newborn hearing screening
NEILS	National Early Intervention Longitudinal Study
NICU	neonatal intensive care unit
NIDCD	National Institute on Deafness and Communication Disorders
NIH	National Institutes of Health
NYS	New York State
OAE	otoacoustic emissions
TC	Total Communication
WHO	World Health Organization

Introduction

MEDICALIZATION, DEAF CHILDREN, AND COCHLEAR IMPLANTS

> *If practices are foregrounded there is no longer a single passive object in the middle, waiting to be seen from the point of view of a seemingly endless series of perspectives. . . . The body, the patient, the disease, the doctor, the technician, the technology: all of these are more than one. More than singular. This begs the question of how they are related.*
>
> • Annemarie Mol, *The Body Multiple*

> *How is any parent to know whether to erase or celebrate a given characteristic?*
>
> • Andrew Solomon, *Far from the Tree*

As I walked up to Jane's home,[1] Halloween decorations hung from the door, and fallen leaves, matted and damp, were strewn all over the ground. Once inside, we made tea and stood in her kitchen, brightly lit by an autumn afternoon sun. Drawings, magnets, and calendars covered the refrigerator. I leaned against the counter as she fed her youngest son, who was sitting in his high chair. We were waiting for her four-year-old daughter, Lucy, to come home on the bus from school. Lucy is deaf and had recently received a cochlear implant (CI). She was in a program that focused on spoken language, relied on speech therapy, and was tailored for children with CIs. As we waited, Jane told me that Lucy was making amazing progress in her speech articulation and comprehension. She was smiling and almost breathless as she talked; it was clear that she was brimming with excitement about it. Our conversation then drifted into how Lucy had been in a different classroom previously, a sign language program, before transferring to her program now. Her new educational placement is called auditory-verbal, and it is part of the same school for deaf children that she was attending. Jane observed that the number of students in the sign language program had quickly dwindled in the past two years. More and more students were getting CIs and transferring to the auditory classroom, she told me. I asked her why she thought enrollment in the sign language program was going down at Lucy's school.

"Nobody's talking about Deaf culture," she said.[2] "With the technology that we're being faced with, it will never have the chance to evolve because it's not big enough. It's such a teeny tiny little culture. We don't talk about it, and the reason we don't is because the majority of parents want their kids to talk."[3] She was pointing out two factors at play: audiologists' and other related professionals' general acceptance of the CI as a tool for learning spoken language *and* the erasure of sign language as a viable option for deaf children. Indeed, Jane celebrated Lucy's speech development. After all, she had engaged in an untold number of hours of care, work, and labor to maintain it, and deeply hoped that Lucy would be able to master it. Yet she also agonized over the pressures she felt to restrict Lucy's language input to only spoken English through the CI. She insisted that Lucy was still a deaf child and sometimes used signs with her. Above all, she was extremely anxious about Lucy's future.

Jane lives near New York City and is one of ten mothers who participated in my study of raising deaf children with CIs. Her experience is embedded within a variety of social changes that have occurred over the past three decades, and her story—as well as the stories of the other families in this study—offers a glimpse into the world of families who are raising deaf children with CIs today. This book identifies the context shaping these families' experiences. One important part of this context is technological change and the powerful influence of medical knowledge (e.g., development of the CI and CI-related clinical and educational programs). Another is the deeply divisive climate surrounding CIs; there has been an enormous amount of controversy over the device. Health policy changes (e.g., newborn hearing screening programs and state intervention services for children with disabilities) have also reshaped the trajectory of parenting a deaf child. Finally, the larger cultural demands on mothers to rely on medical interventions in child rearing also affect the way parents respond to their child's deafness.

Using observations and in-depth interviews, in this book I describe the institutional structure and culture of CI-related interventions for deaf children. Some of my fieldwork was conducted in a CI clinic and depicts the organization of and particular therapeutic culture found in a CI center. Other times my fieldwork took place with parents in their homes, and as they took their child to school or attended parent events. In the following chapters is a detailed description of families' experiences of the therapeutic culture surrounding the process of learning their child was deaf, deciding to get the child a CI, and navigating the multiyear process that precedes and follows surgery. Perhaps the most crucial aspect of implantation is how much time it takes; it is not a onetime surgical event. Time is needed to prepare for the CI and to learn to use it. For parents, this long process is fraught with grief and hope, persistence and frustration, and conflicting information and emotions. For example, they are aware of the political battles and controversy over the use of CIs, but such debates can be alienating because parents' concerns differ strikingly from the political battles that play out over the device. In addition, professional advice for deaf children changes over time; different professionals give different advice, parents draw from their own gut instincts, and yet sometimes their "nonmedical" knowledge is minimized in the context of the clinic. And then there is the unpredictability of the outcome: parents may struggle for months or even years before observing spoken language in their child, and mastery does not always come.[4] Thus, there is a continual tension between their present efforts and their hopes for their child's future. In general, clinical research is not too concerned with the struggles that Jane and other mothers described to me as they learned their child was deaf, made treatment decisions, and raised a deaf child with a CI. This book documents those very struggles.

The families' experiences in my study were powerfully affected by what I call *ambivalent medicalization*. Medicalization is an important sociological concept that describes how we understand conditions through medical language, medical thinking, or a medical framework.

Academic theories and studies of medicalization tend to be concerned with how medicalization has power over people, especially through affecting their behaviors and influencing us to see differences between people not as neutral variations but as medical problems. However, studies have also shown that people experience medicalization in ambivalent ways (e.g., Blum 2011). On one level, I use the term *ambivalent medicalization* to include the sometimes ambivalent feelings that mothers in this study experienced (although more often than not, they were socialized to *not* be ambivalent). But on a deeper level, I use the phrase in conversation with theories of medicalization. Overall, studies of medicalization emphasize its power and therefore the lack of agency that individuals have when they encounter it. But *ambivalent medicalization* emphasizes that individuals are both empowered by *and* surrendering to the process of medicalization. In this book, I argue that medicalization and biotechnological tools create new ways of being, and that this has, simultaneously, both good aspects and bad aspects. That is, ambivalence is not just what people might experience as a result of medicalization, but rather it is characteristic of the process itself. Some things are gained while others are lost, or as Riessman (1983) put it when writing about medicalization and women's lives, it is a "double-edged sword."

Ambivalent medicalization is a concept I develop here to account for all of this, and it has broad application and a number of features. Most importantly, it is a tool to close the gap between theorizing medicalization and the humanistic enterprise of accounting for the experiences of individuals in society experiencing it. First, it recognizes the sheer triumphs of medicine, the technoscientific feats we have available to us, and the possibilities and relief these options often provide to people with all kinds of diagnoses, from minor ailments to those involving death, angst, and pain. But it also acknowledges that there is an underbelly to medicalization. It often requires individuals to submit themselves to medical thinking and clinical logic that—despite the hope offered—can be fraught and difficult in day-to-day living. As I will show, medical interventions can have real benefit, and it soothes families to have a plan of care in the face of a troubling diagnosis. Sometimes families actively resist the pressures that come with these interventions, but most often families are being enculturated into medical thinking, and in the process they are told to accept some sets of resources and reject others. The concept of ambivalent medicalization tries to capture both the gains and the losses that come with medical and technological advancement; for while some things may be attained through medicalization, other ways of being or communities may be lost. Overall, it *recognizes* the capability of medicine and the relief it provides, *attends to* the accompanying profound social and ethical implications, *accounts for* the labor that medicalization often requires of patients and caregivers, and *grapples with* the fact that medicine may not always yield the best—or even the intended—results.

Through interviews and observations, I show how ambivalent medicalization works as a set of tensions, a nexus of certain trajectories and conflicting points of view. First, tensions over whether deafness is a medical problem in the first place play out in parents' lives. But ultimately almost all parents tend to adopt a certain "script" about deafness. The different scripts for narrating and giving meaning to what it means to be deaf serve as a backdrop for understanding the many issues surrounding deaf children and CIs, as well as contribute to the ambivalence experienced by parents. Second, the availability of CI technology, which is a neuroprosthetic, puts the development of the brain at the center of all interventions. The centrality of the brain is the most significant feature of the scripts regarding deafness that parents adopt. Third, in the past forty years or so, responsibility for the care of children with disabilities has shifted from institutions to the home. I show the consequences of this shift for motherhood, since it co-occurred with the rise of medicalization and mapped directly onto gendered patterns of care work. As a result, tremendous pressures are put on mothers of children with disabilities to proactively engage in medical interventions to "overcome" them. Finally, socioeconomic status (SES) and cultural background play a significant role in which mothers do this work and how they are seen by clinicians.

Contextualizing Deafness: Different Scripts Framing the "Problem"

Professionals in implantation and many in the general public most likely agree that deafness is a disability[5] and should be fixed. For example, audiologists work to measure the amount and kind of hearing loss a child has, recommend tools like hearing aids and CIs to correct or mitigate it, and coordinate with various cooperating intervention programs to support those goals. Parents are expected to bring their child to the clinic as needed, maintain the device's functioning, and solicit speech therapy—and other services—for their child. There is little controversy among professionals over these basic expectations. These accepted norms around treating deaf children and the assumption that they should learn to hear and speak are, however, also a product of a larger culture.

One way of explaining this script about the meaning of deafness is to go back to the sociological concept of medicalization. In general today, medical knowledge has come to dominate how we talk about, understand, and respond to many things in our society, our lives, and our bodies. Irving Kenneth Zola's (1972) work originally conceptualized medicalization and argued that it was the expansion of powerful medical knowledge into ever more areas of life. Labeling something a medical problem is also a key feature of medicalization (Brown 1995; Schur 1972). Peter Conrad gives a working definition of medicalization as "defining a problem in medical terms, using medical language to describe a problem, adopting a medical framework to understand a problem, or using a medical intervention to 'treat' it" (1992, 211). As Conrad sees it, the essence of medicalization is definitional, although this does not mean it is static. For example, he points out a variety of conditions that were once seen as medical problems but have now largely been de-medicalized, such as homosexuality. This is partly because of the gay and lesbian movement's fight to remove homosexuality from the purview of medicine, and thus it was removed from the Diagnostic and Statistical Manual of Mental Disorders (DSM) in 1973 (Conrad 2007). Still, at the same time, other once "natural" aspects of life—such as death, birth, and pregnancy—became more medicalized. Determining whether something is a medical problem is not the goal of sociological work. Rather, the aim is to show how something is understood as a medical problem, how this norm is accepted or challenged, and the consequences this has on people's lives.

To that end, there is another script for deafness. In the 1970s, deaf people started distinguishing between the lowercase *deaf,* which describes one's audiological or hearing status, and the capitalized version of *Deaf,* which refers to an identity of being culturally Deaf (Padden and Humphries 1990). The beginnings of the Deaf cultural movement in the United States coincided with a variety of new social movements in the latter half of the twentieth century, such as the civil, gay and lesbian, women's, and disability rights movements. Many of these movements championed a similar way of thinking about difference. They took what society deemed a stigmatizing trait—often framed as a disease or a condition—as, instead, the basis for an identity, community, and culture to be celebrated. The Deaf cultural movement argued that deafness is not inherently pathological but has been collectively made or constructed as such through history, the language we use to describe or label people (e.g., medical language), and systems of power that privilege some people over others. Today there is extensive literature published on Deaf culture, a culture with its own history, language, and way of seeing the world. Much of this work argues that being Deaf is primarily about the use of sign language to communicate and that cultivating Deaf culture is of paramount importance as a way to offer a counter-discourse to that of medicalization. Central to the Deaf critique is that they are a linguistic minority, and, as the following chapters will show, the politics of language are at the center of any debate over implantation.[6]

During this same time period of new social movements and the expansion of Deaf culture, the CI was being developed. As Stuart Blume explains in his book *The Artificial Ear* (2010), during the 1980s and 1990s, the CI moved from being an experimental innovation to being clinically feasible. Early on in its development, while still in the experimental phase, it was used in adults, who reported various levels of benefit from the device. Blume (2010) and Mills (2012) both show how the adult CI users collaborated with the CI designers. But as the research side of CIs gained attention, the device's broader viability needed further testing. Clinics needed to identify more candidates to test these developments and improve upon them. Garnering broader participation required, however, a public message that resonated with potential CI candidates. And here there was a problem. The market for CIs in adults "grew far more slowly than had been anticipated. . . . Deaf people were not coming forward in anything like the numbers anticipated by professionals and manufacturers" (Blume 1997, 38). Indeed, Deaf adults—who were part of Deaf culture—simply did not view their deafness "in the same terms as medical and audiological professionals: as a loss of hearing" (Blume 1997, 39). To them, nothing needed to be fixed because their deafness was an identity, their language was American Sign Language (ASL), and they were part of a culture. Thus, because of these vastly different scripts around deafness, adult CI patients turned out to be a less accessible market than they had hoped.

While Deaf adults failed to be as interested in CIs as professionals had hoped, changes in the identification and diagnosis of deaf children were making a new market for CIs possible. Various changes in legislation and health policy contributed to deaf children being identified much earlier, which I explain further in the following chapter. Then in 1995, the National Institutes of Health (NIH) convened a consensus meeting of specialists in fields related to CI, as well as representatives from the public. After reading reports, studying data, and hearing testimony, members of the consensus group concluded that although the success of implantation was highly variable and these variations were unexplained, the data also showed a trend indicating that the shorter the duration of deafness, the better the individual's performance (e.g., ability to understand and/or produce speech) with a CI (National Institutes of Health 1995). It was this fact that resonated most with the group. By 1998, despite protests by the Deaf community and testimony by some of its members, the age of implantation was lowered to eighteen months. By 2000, it was lowered further by the Food and Drug Administration (FDA), to twelve months.

These two strands of history—the rise of the Deaf cultural movement and the development of the CI—barreled toward each other on simultaneous timelines. Because of this, deep controversy has surrounded CIs since their inception. Many who subscribe to the antimedicalization script about deafness are opposed to CIs. I call this antimedicalization view of deafness and subsequent opposition to CIs the *Deaf critique*. Those arguing against CIs still do so today by citing the value of Deaf culture and usually depend on an identity politics that, at its core, emphasizes Deaf culture and the use of sign language as a kind of biodiversity (Bauman and Murray 2009). This view asserts the legitimacy of what Deaf scholar Ben Bahan (2008) calls a "visual way of being" and values the shared experiences of Deaf culture and, most importantly, the use of sign language.

By contrast, the view that deafness is a medical problem is the crux of clinical approaches to deafness and a starting point from which clinical interventions begin. Because more than 90 percent of deaf children have hearing parents and deafness in children is now identified much sooner, children under the age of six (the demographic category used by the National Institute on Deafness and Communications Disorders [NIDCD]) are the fastest-growing group of CI recipients (NIDCD 2015). That makes hearing parents the primary consumers of CIs, not d/Deaf persons. Even though some do not advocate for such a strict divide over CIs and Deaf culture (e.g., Woodcock 2001) and there has been some documentation that the controversy over CIs has subsided somewhat, "opposition to pediatric implantation among certain members of the Deaf community continues unabated" (Christiansen and Leigh 2010, 47). It is for these reasons that this book focuses exclusively on pediatric implantation.

This book focuses on the lives of those parents who opt for implantation because, despite the Deaf cultural movement and its arguments against implantation, the CI has shifted from the peripheries of innovation to become the most advanced and commonly used neuroprosthetic. Today, the CI is the world's largest medical device market, and, as of 2012, the FDA reports that approximately 58,000 adults and 38,000 children in the United States have been implanted (NIDCD 2015). The most rapid growth in the industry has occurred during the past ten years.[7] Cochlear, the largest CI manufacturer, reported that "fifty percent of people implanted with a CI received it in the last five years, reflecting the exponential growth characterizing this intervention" (Cochlear Americas Corporation 2008, 1). Yet even as the Deaf community laments the expansion of implantation, those who advocate for CIs sound alarms about its underutilization (e.g., Sorkin 2013), illustrating that there are vastly differing viewpoints on the device.

Rather than focus on trying to catalog these viewpoints, I show how when a child is diagnosed as deaf, parents find themselves in the middle of two very different scripts on deafness and a fierce battle over language politics. The Deaf critique of CIs comes from

Deaf experience and sees the problem as society's intolerance of difference. But deafness is not often passed on genetically, so often, parents do not have this Deaf experience to draw from. Indeed, it is genetic in less than 10 percent of the cases, which is why more than 90 percent of deaf children have hearing parents (NIDCD 2015). Thus, because medical intervention is so heavily relied on in contemporary culture, parents primarily come to understand deafness via medical institutions that provide the diagnosis. This presents a starkly different script than the Deaf perspective does, but parents must grapple with their own desires, expectations, and hopes for their child. Previous studies of implantation have shown a strong correlation between parents' status as hearing and their desire for their child to develop spoken language. And that desire for the acquisition of spoken language has been documented as the most significant factor in the decision to implant.[8]

But how do these ways of seeing deafness and these debates over which script to adopt figure into this book? After spending time with parents of deaf children with CIs, it became clear to me that these two polarized narratives about deafness, which are also seen in the public discourse about implantation, are inadequate. Parents do tend to adopt the medicalized script of deafness, but they do so in a particular context that funnels them toward this script. Along the way, they may also experience moments of ambivalence. In chapter 2, I show how the initial adoption of the medicalized script happens through interactions and institutional protocols that I call *anticipatory structures*. Many of the professionals working in fields related to implantation who were a part of this study actively encourage, and are complicit in, maintaining strict divides between these competing scripts of deafness. In the following chapters, I will show how parents are socialized into this medicalized script of deafness, describe the language debates between speech and ASL, and examine the increasing discourse about the brain.

Biotechnology and Disability

The simultaneous development of the CI and the Deaf cultural movement illustrates a paradox that has broader implications beyond just implantation: Public discourse about diversity and accepting difference has steadily increased in the United States, yet the tools of science and medicine that are used to treat various conditions (which may very well be included under the umbrella of diversity) have grown more sophisticated and available. We can now engage in attempts to treat, find relief, look or feel more "ideal," or pass on traits of normalcy—which have social value—to our children. The families in this book made decisions that simultaneously reflected dominant norms regarding the use of biotechnologies *and* regarding disability. These two sets of norms cannot be disentangled from one another, yet both tend to be seen as "common sense" and often go unnoticed.

For example, we often think of science, technology, and medicine as neutral or objective endeavors and authoritative explanations. However, science and technology studies (STS) scholars have shown that rather than emerging from an objective or "natural" truth, science is collectively made through interactions. Sociologists of knowledge and STS scholars have long said that science and medicine are social and that technological artifacts are not neutral tools but rather are embedded in and inscribed with social and political relations.[9] More specifically, STS scholars Clarke and colleagues (2010) argue that medicine is now so interwoven with technologies that the process of medicalization described by sociologists has been fundamentally altered into something else: biomedicalization. Medicalization is more about control over disease processes, while biomedicalization is about transformation through technoscientific means (Clarke and Shim 2011).

This technological shift and emergence of biomedicalization have a number of consequences, one of which is that technologies and new scientific knowledge change how we understand conditions and construct those conditions' meanings; that is, they are intertwined as sociotechnical systems. Increasingly, the effects of technologies are being interwoven into sociological examinations of medicine (Casper and Morrison 2010), and as I will show in chapter 4, the technology of CIs contributes to an important shift in the definition of deafness. Because the CI is a neuroprosthetic device, deafness has come to be redefined from a sensory (hearing) loss to a neurological (processing) problem. One result of this is that the CI is then constructed as merely a tool providing access to the brain, which is the site of the "real" treatments. These "real" treatments are the long-term therapeutic endeavors parents engage in, which are focused on neurological training to *transform* the child's brain into one that functions as much like a hearing brain as possible. In chapter 4 I show the profound impact of this focus on the brain on how deafness is medicalized and what families who opt for the CI experience are expected to do.

While the development and expansion of technologies transformed the process of medicalization into biomedicalization, norms around disability also shifted and were critiqued.[10] This is due in part to the work of the disability rights movement and the emergence of the field of disability studies. Like the social movements I mentioned earlier, the disability rights movement argues for seeing disability not as a medical problem but rather as stigmatization or a social problem.[11] Two models are put forth for analyzing disability, the medical model and the social model. In the medical model, a person's disability is an individual pathology, socially undesirable, and to be cured or mitigated using medical intervention. The social model, however, distinguishes between "impairment" (the physical condition of the body) and "disability" (the social attitudes toward that body). Like technologies, bodies are inscribed with social meanings, and these meanings are made collectively. In the social model, it is social barriers, social attitudes, and unwillingness to accommodate different types of bodies and needs that produce disability, *not* only the body or its condition. This model led to a counterdiscourse wherein disability became a point of pride, provided a link to a broader disability culture, and offered opportunity for political action.[12]

These models of disability ultimately critique the ideology of ableism. Ableism has been defined by a variety of scholars, but here I rely on Campbell's definition as "a network of beliefs, processes, and practices that produces a particular kind of self and body (the corporeal standard) that is projected as the perfect, species-typical and therefore essential and fully human. Disability then is cast as a diminished state of being human" (2009, 5). Interestingly, the Deaf critique of implantation clearly echoes the social model of disability and the description of ableism, although the relationship between Deaf studies and disability studies is a tenuous one at best. Indeed, some readers may even bristle at the categorization of deafness as a disability in this book. Scholars in disability studies and the disability rights movement often include the Deaf movement and "claim" the d/Deaf community as part of their own constituency, but the Deaf community does not usually reciprocate and often rejects association with disability in favor of Deaf-specific models of culture and community around deafness.[13] This is largely because of the presence of language that accompanies deafness and how this profoundly shapes the particular experiences that Deaf people have that are different from other disabilities. One example of tensions between the Deaf and disability movements is found in Shapiro's historical account of the disability rights movement, where he notes that for Deaf persons "one of the first great victories of the disability rights movement—the mainstream education law—was a threat. It led to cuts in public funding for segregated deaf schools" (1994, 100).

In this book, I approach deafness as a particular case of disability and see the disavowal of disability on the part of the Deaf community as merely reproducing discriminatory attitudes, in short as ableist. Delving into the particularities of the Deaf experience is indeed important, but it need not require distancing from disability, and in fact may be made all the more richer in dialogue with it. Why is the goal of Deaf studies and the Deaf cultural movement not to advocate for the flourishing of all communities of those whose bodies—and brains—fall outside the norm? One overlap can be found in studies of autism and the relatively recent turn toward claims of neurodiversity. The main claim of the neurodiversity movement is again similar to the Deaf critique of CIs; autism is restated as a form of brain architecture that is simply another human variation, or an example of neurodiversity, and should not be cured or otherwise fixed.[14] Similar to the tensions that are explored in this book, the arguments over and meanings of autism differ across stakeholders—from people within the neurodiversity movement and autism community, to professionals who espouse (and policies that support) medical interventions, to families making choices about which services and treatments to undertake. Furthermore, deaf persons sometimes have other disabilities as well. But overall, I see the tendency of Deaf persons to eschew disability and the disability community as anathema to the work of both movements.

In sum, in the past few decades, technology changed medicalization into biomedicalization, and the disability rights movement began to frame disability as human diversity. These strands of thinking in

STS, sociology, and Deaf/disability studies tend to share a critical view of medicalization and biotechnologies. As a result, scholarship in STS is often accused of being antiscience or antitechnology (Bauchspies, Croissant, and Restivo 2006). In addition, Conrad (2007) has asserted that the sociological concept of medicalization describes a neutral process, but nevertheless it is typically constructed as negative (Parens 2011), as most sociological studies tend toward focusing on the power of medicalization over us and building a "symbolic case against medical hegemony" (Timmermans and Berg 2003, 101). Lastly, the social model of disability and the Deaf critique are both explicitly antimedicalization. Despite accusations of being antiscience or antitechnology, however, these fields are not. What they share is a commitment to nuance, to critical thinking about the ways we use technologies and the ways we use scientific and medical knowledge, which is what draws them together so well. This is also why I emphasize the ambivalence of medicalization. It is not a neutral process; technology is not neutral, bodies are seen as in need of transformation (especially disabled ones), and the social structures where these processes take place are cultural sites with norms and values.

The politics of disability and the consequences of biotechnologies have certainly been examined together before.[15] However, the patterns I observe make explicit that setting up stark choices between models of disabilities (in this case narratives of deafness) and decisions about whether to endorse or critique medicalization are woefully inadequate. Others have already noted the limitations of the social model.[16] As Shakespeare and Watson state, "People are disabled both by social barriers and by their bodies. This is straightforward and uncontroversial" (2001, 23). An excellent example of how this has already been studied in the family context is Landsman's (2008) research showing that mothers of children with disabilities vacillate between these models without total allegiance to one or the other.[17] Meanwhile, in public and academic debates on CIs specifically, many have disagreed for decades about which "side" is right; these tensions have been well documented in mass media (Davey 2011; Boggs 2010; Solomon 1994; Aronson 2001). In contrast, this book shows that there is a striking difference between theories of medicalization, public debates over CIs, and families' concerns as they raise a deaf child.

This gap between the debates over CIs and families' experiences is far more aptly described by the concept of ambivalent medicalization I outlined above. Previously, Parens suggested that we "try to articulate the difference between good and bad forms of [medicalization]" (2011, 2), but ambivalent medicalization takes this idea a step further, allowing for both good and bad consequences of medicalization to coexist and recognizes ambivalence as characteristic of the process itself. It also allows for the politics of disability and the politics of medicine to be considered together. For example, Kafer (2013) developed a political/relational model of disability, and this model neither embraces nor rejects medical intervention but instead recognizes the simultaneity of desires for cure or mitigation *and* being allied with the social and political struggles of people with disabilities. By deploying *ambivalent medicalization* then, I seek to capture the nuances of the daily, lived realities of families caring for a member with a disability and address the inadequacy of the bifurcated models of disability and ways of thinking about medicalization.

Emphasizing the Family Context

Pediatric implantation, and any medical intervention for that matter, cannot be disentangled from family life. Today, social practices and policies regarding disability are very different from before. For example, the disability rights movement fought to end institutionalization, the cultural practice of putting a child born with a disability into an institution for the duration of his or her life. As a result, the site of care for children with disabilities largely shifted to individual families in the United States starting in the latter half of the twentieth century. This shift overlapped with the emergence of biomedicalization and parents increasingly taking on a collapsed parent/patient role in efforts to care for their disabled child. Together these shifts signal that caring for a child with a disability is a private matter. Children are seen as "private property, and their economic and social burden is not shared" (Rothman 1993, 8). Engagement in curative techniques is the family's responsibility.

This responsibility is highly gendered. It is well documented that child-rearing duties primarily fall on women, with a "persistent ideology about the gendered nature of care work" (Herd and Meyer

2002, 2). This is no different when it comes to children with disabilities; it is most likely the mother who will be in the collapsed role of parent/patient. Indeed, in my study it was almost entirely the mothers who were in the clinic attending appointments, and who, as in previous studies of mothers of children with disability, "provided the majority of the therapeutic care, and were held responsible by professionals for that care" (Leiter 2004, 839). This is aptly illustrated in the following chapters, as clinical staff consistently refer to the parents involved in implantation as "the moms." And so this book is also a feminist endeavor to include women's voices in the knowledge production related to CIs.[18]

Motherhood intersects with biomedicalization in particular ways. For example, one feature of biomedicalization is how "the key site of responsibility shifts from the professional physician/provider to include collaboration with or reliance upon the individual patient/user/consumer" (Clarke et al. 2010, 65). Also, biomedicalization illuminates that health is a moral obligation that requires self-transformation (Metzl and Kirkland 2010; Rose 2006). In the following chapters, I show how mothers' obligations to transformation and collaboration are explicitly, consistently, and strategically invoked in the clinics, the supporting institutions, and the community of parents of children with CIs. The message is clear: a "good" mother will do the long-term care work associated with implantation if her child is deaf.

Scientific motherhood, or "the belief that women require expert scientific and medical advice to raise their children healthfully" (Apple 1995, 161), undergirds CI-related interventions in children. But the ethos of scientific motherhood is often seen as common sense, rather than a historically and socially constructed belief. Indeed, the therapeutic duties mothers are expected to follow are dictated by health and intervention policy in the United States, which is founded on scientific knowledge (Leiter 2004). In this book, I show how today this duty is primarily defined by a focus on the brain and requires a commitment by mothers to capitalize on neuroplasticity.[19] As a result, mothers are socialized into the idea that a regimen of "neuronal fitness" for her child is the key to individual success (Pitts-Taylor 2010).

Notably, this shifts responsibility *from the device to the individual* and, in the case of pediatric implantation, onto the mother. While on the surface we celebrate the CI as a technological triumph, we are in reality demanding more and more invisible labor on the part of mothers to achieve its successes and blaming mothers if it does not work. As we increasingly depend on neuroscience as a trope of explanation, the site of labor for mothers then goes deeper into the body, down to the wiring of the brain. Thus, what unfolds in the following chapters is an exploration of the neuroscientific narrative's framing of families' individual responsibility to ensure the normalcy of their child. This convergence of scientific knowledge and individual, private responsibility assigned to mothers illustrates Dána-Ain Davis's observations that when accounting for women's experiences, "particular domains of expectation unfold especially in relation to neoliberalism" (2013, 25).

While scientific motherhood is encouraged by institutions involved in CI-related interventions, socioeconomic status and cultural background affect the extent to which mothers engage with and embrace the role. Numerous studies confirm the positive correlation between "high levels of family involvement" in interventions and outcomes in childhood development (e.g., Conger, Conger, and Martin 2010). Lower-class families have less social capital, fewer resources, and less time to participate in the ongoing intervention strategies. Indeed, class disparities have also been identified in pediatric implantation.[20] In chapter 3, I show that disparities in pediatric implantation occur through two channels related to class and cultural background. The first is related to the biases professionals in implantation have regarding socioeconomic status and diverse cultural backgrounds. For example, audiologists are less likely to deem children of lower socioeconomic status and immigrant families as appropriate CI candidates (Kirkham et al. 2009). The second channel echoes Lareau's (2003) observations that upper- and middle-class families adhere to a different parenting style than working- and lower-class families. These parenting styles have different attitudes toward institutions; in white and/or middle-class homes, Lareau argues, the "high intervention" parenting style reflects the values of formal institutions. This creates what Lareau calls a more "seamless overlap" between institutions and the home.

The Research Setting and Study Participants

This book is based on a multisited ethnography. Over the course of approximately six months, I conducted fieldwork in a CI clinic two or three times per week; interviewed ten parents in their homes; and observed school programs, parent support groups, and other parent events. To find a site where I could conduct clinical observations, I visited the website of Cochlear, one of the largest manufacturers of CIs. Clinics where one can obtain a CI were listed by area. I then e-mailed the director of every center listed in the New York City metropolitan area. The director at one of the clinics, which I will call New York General, or NYG, replied. After an initial meeting about the goals of my research, she approved my study idea.[21]

A few days later, the director of the clinic at NYG, Sharon, asked me to give a presentation during lunch to the entire staff of the center, about twenty people. I explained my interest in understanding how implantation was socially and institutionally organized and my focus on the parent experience. I also explained that I was a sociologist who would probably be "lurking about," and that they should go about their usual professional duties. Indeed, I often simply accompanied audiologists throughout their day.

The NYG CI clinic is housed in a large hospital system and is part of a hearing and speech center that treats patients with a variety of audiological conditions. The clinic participants were the audiologists, surgeon, speech therapists, and social worker who work in the area of pediatric implantation, as well as the parents who interact with them. In the interest of protecting those involved, I have changed the names of all participants and the name of the clinic where I conducted fieldwork.

NYG provided me with an identification badge similar to those worn by all employees. It bore the title "Volunteer." I was assigned to the hearing and speech center and was given a desk and computer in the cubicle next to the secretary to the CI surgeon. I dressed professionally at all times. I had access to the scheduling software, the patient database, and patient histories (which I accessed only after parents signed a consent form allowing me to do so). I had permission to sit in on staff meetings and CI team meetings, and to talk to center staff and be present during daily routines. The day-to-day operations of the clinic include an array of services, but the focus of this book is on the specific processes and appointments related to infant CI recipients, such as evaluation, candidacy, surgery, and aspects of long-term follow-up care.

I also traveled to parents' homes to conduct interviews. The inclusion criterion for the parents in my study was that they had a child under the age of six who had been, or was about to be, implanted. I chose this age group to align with the NIDCD statistics showing this age range as the fastest-growing demographic of CI recipients. I recruited parents through the clinic and through parent events such as support groups, informal dinners, and advocacy trainings arranged by local hearing loss–related organizations. (For example, CI corporation Cochlear sent representatives into schools to run parent trainings on how to best use the device.) Because almost all of the parents in this study were recruited at NYG, I was able to interview audiologists and parents about the same clinical encounters and get multiple perspectives on one interaction.

The range of time since implantation varied for the children of the parents I interviewed. Some had been implanted thirteen years prior to my interviews with their parents, while others underwent surgery during my time at the clinic. More uniform was the age of the child at the time of surgery: averaging two years. Candidacy determination, pre-implantation appointments, surgery, surgical follow-ups, and multiple types of long-term care take place over a period of years. For a variety of reasons, including the fact that implantation is a multiyear procedure, I could not follow parents through the entire process. Thus, the participants in my research were all at varying points of the implantation and rehabilitation process and were typically moving from one phase to the next. To give an impression of the process, I offer stories from different parents at different stages. During observations and interviews, I used a digital voice recorder to record interactions and always had a notebook with me, where I wrote down field notes. All field notes, recordings, and interviews were transcribed and analyzed line by line for recurrent themes.

One of the limitations of this study is that the interactions described here are those of only one clinic, and therefore may not be indicative of national trends. Currently, there is no one nationally defined protocol for CI clinics, although they typically follow U.S. Food and Drug Administration guidelines and CI manufacturers' recommendations (Bradham, Snell, and Haynes 2009). In fact, the variation across clinics is attributable to a lack of established, national-level best practices. The American Academy of Audiology (AAA) has published advice about CIs, and the American Speech-Language-Hearing Association (ASHA) has published preferred practice patterns. However, those two resources "provide comprehensive discussion but not enough specificity to be considered as best clinical practices" (Sorkin 2013, S9). As a result, implantation practices and CI information dissemination vary considerably within and across states (Sorkin 2013). Since I did not survey every other clinic in the country, I do not know to what extent the practices at NYG resemble practices at other clinics. Future comparative studies would be beneficial. Nevertheless, I hope that this study can give a picture of what some families experience and perhaps suggest additional important areas of inquiry when establishing such standards of practice.

How a CI clinic operates depends on its resources. The extent to which resources are allocated to CI programs depends on a number of factors, such as whether the clinic serves a major metropolitan area and its proximity to educational programs. NYG has a particularly rich supply of resources because it is located in one of the largest health care systems in New York State and is in proximity to a number of deaf schools and programs in the New York City area. As such, it is most likely to be generalizable to clinics in larger cities with available CI community resources, such as schools.

A second limitation of this study regards the sample size and uniformity. The parents I had access to are those whose children had received a diagnosis of deafness and who were getting, or already had, a CI. I do not know what happened to parents of deaf children who did not opt for a CI, nor did I ever witness appointments where parents stated that they did not want a CI. Because families who did not get the CI were not in the clinic regularly, I did not have exposure to them. When I asked the audiologists I worked with about this demographic that was hidden to me, they told me that typically it was the small percentage of deaf children who had deaf parents that did not get the CI and simply did not need to come to the clinic.

This is not to say that deaf parents do not ever get CIs for their deaf children (because they sometimes do), but rather that this limitation of the study reflects the experiences of the audiologists at this clinic and the clinic's trends in implantation. As I take up in chapter 3, children receiving CIs are members of predominantly white, middle-class families (e.g., Johnson 2006; Boss et al. 2011; Hyde and Power 2006). Thus, my sample of primarily white, middle-class mothers who frequented NYG reflects these trends. I acknowledge that there are variations in CI experiences beyond my sample, but the patterns I found in the overall structure and culture of implantation offer a starting point for more broadly understanding the social aspects of pediatric implantation.

Dilemmas in Fieldwork

Ethnography is a method of making a social world visible through deep description, attention to microinteractions, and interpretation of the processes of meaning making. All ethnographers must wrestle with their positions in the field; any ethnography is by definition full of awkward social relations, and it is the job of the ethnographer to navigate these (Hume and Mulcock 2004). But reflexivity—the awareness of both the ways we shape the environment in which we conduct fieldwork and the ways in which our own experiences and biases influence interpretation of data—takes a particular shape when researching controversial topics. One must think carefully about assumptions and critically about any claim of neutrality.

In his classic essay "Whose Side Are We On?" (1967), Howard Becker argues that the sociological researcher can, in fact, never be neutral. Indeed, I have a background in the Deaf community as a friend, advocate, and professional sign language interpreter for more than twenty years. This experience shaped the assumptions I had going into this research project, but it did not prevent me from seeing the complexities of implantation. This was partly because I borrowed some of my methodological strategy from grounded theory, a type of ethnography where one goes into the field without a specific hypothesis to prove or disprove (Strauss and Corbin 1990; Charmaz 2006). Because I had little idea of what to expect when entering the clinic, grounded theory gave me an open orientation wherein I adopted the tactic of listening to and recording everything. In addition, I adopted the strategy of "situating knowledge" from feminist theorist Donna Haraway. This required a commitment to mapping the tensions around a particular subject, understanding the views of different stakeholders involved, and seeing things from different perspectives (Haraway 1988).

My overall strategy was not to prove one side of the debate right or wrong but instead to show the people I studied and what meanings the CI had for them. I also wanted to depict the study participants as complex individuals. Parents, for example, at times adhered to what they were told by persons in positions of authority and at other times carved out their own systems of meaning. No matter what their role, all participants struggled to do what they felt was best. But what is "best" is socially constructed. So in this book, I emphasize and critique the social, cultural, and institutional structures shaping these beliefs and actions in the first place, rather than critiquing individuals or their choices.

I cannot know how or if my background affected the way study participants interacted with me or with others while I observed. Although I disclosed my background in the Deaf community to the CI center director before the study commenced, and she still welcomed me, I do not know to what extent that influenced my observations and interviews in the clinic. I also had to decide if and when to disclose this background to parents. In general, I did not bring up the controversy over CIs, although participants in the study sometimes did, and I never gave an opinion on the matter.

I am acutely aware that readers will map their own cultural framework or allegiance onto the material presented in this book. Although I try to emphasize the complexity of pediatric implantation and what these families experience, it is apparent that I am critical of the process of medicalization. But I am equally critical of some of the arguments from the Deaf critique; that is, this book is grounded in the data my fieldwork generated, driven by an analysis of the patterns I observed, and as a result presents a nuanced view on CIs that does not fully accept the terms of either side of the debate.

This is why I chose to title this book *Made to Hear*. With the title, I wanted to set a tone that reflected ambivalent medicalization—that is, encouraged shifting, various, and simultaneous meanings. I thought of myself at work as an interpreter, trying to interpret or translate the meaning of the English word *made*. When you interpret from spoken language to American Sign Language, the actual English word does not matter or directly translate; what you interpret is the concept that roots it. The English word *made* has multiple meanings, and in ASL there is a different sign for each of these. For example, the title of the book could be referring to *made* as in individuals rendered able to hear through the facilitation of a device. Or it could be *made* as in being forced or compelled to engage in the act of hearing. Or it could be *made* as in a belief that we are created, as humans, to hear. You have the option to interpret *made* in any way you want and perhaps in multiple ways at once.

The Chapters Ahead

Chapter 1 emphasizes that implantation is a long-term process. I identify the five stages of implantation, show how the children are diagnosed, and how the parents are inculcated into a medicalized script of deafness through microinteractions and anticipatory structures. Chapter 2 is a continuation of this theme but focuses on the emotional supports available to parents as they move into the second stage of implantation. In this stage, each state initiates Early Intervention (EI) services, a program that supports children with disabilities from birth until the age of three. Chapter 3 details how children move into the next stage of implantation: candidacy. In this chapter, I show how audiologists' assumptions about the impact of a family's class position and cultural background influence parents' adherence to clinical recommendations. Chapter 4 focuses on the central role that neurological discourse takes throughout all of the implantation stages. Chapter 5 further elucidates the interinstitutional connections between clinics, CI companies, and schools. It shows that certain school programs have become an extension of the clinic, especially as new education industries arise specifically to meet the needs of implanted students. In the conclusion, I revisit the patterns I found regarding clinical socialization strategies of mothers, the influence of class on implantation, the role of neuroscience, and the expanding markets in the CI-related education sector. In doing so, I make the case for the utility of ambivalent medicalization as a concept that better explains the patterns described in this book. I also identify questions we should be asking about implantation, suggest critical alliances that might help address issues in implantation, and comment on the continuing controversy over the device and schism between implantation and Deaf culture.

Much of the previous literature that explores social aspects of CIs explains why people have such differing opinions and often defends one side of the debate or the other. By contrast, this book offers a balanced look at both sides of the issue, does not fully accept the terms of either side of the debate, and uses empirical data to construct an analysis of the social aspects of implantation. Families' experiences and concerns are far different from the political arguments. Indeed, the family is a key social site where meanings of deafness and disability are assigned but also perhaps rewritten (Rapp and Ginsburg 2001). What does the medicalization of deafness in families look like, and what meanings of deafness propel these children into the future? By slowing down and pulling apart the moments in these families' experiences, implantation is revealed to be long and arduous. What emerges is an ambivalent medicalization, one that both demands work of parents and provides relief, allows for accepting some aspects of one script and rejecting others, gaining some things while losing still others.

In our society, the Deaf cultural script of deafness tends to be dismissed as merely ideological, and professionals in the fields related to implantation tend to view both the Deaf critique and the use of sign language as harmful. Meanwhile, opting for biotechnologies and medical intervention is generally held up as the correct choice, and mothers are expected to do the care labor that accompanies this medicalization, though not all have the socioeconomic resources to do so. Over the following chapters, I make clear that social factors, not technological prowess, may predict the outcome.

Medical knowledge about deafness and the deployment of CI technology is socially organized and historically situated. The real intervention of this book is to show how the medicalized script of deafness and scientific claims about the brain used by audiologists and other professionals in implantation are as equally cultural as Deaf cultural claims about what it means to be deaf. As in other studies of controversial topics (e.g., Scott, Richards, and Martin 1990), professionals and educators, armed with scientific knowledge, refer to opponents' claims as irrational, unscientific, and thus merely "ideological." But the production, consumption, and characteristics of medical knowledge regarding deafness are also collectively made—that is, medical or scientific claims are also social claims. This book makes that process visible and by extension also argues that it can be collectively made differently. The two points that Jane alluded to that afternoon in her kitchen—that the CI is now largely embraced as a tool for treating deaf children and that sign language is being erased as a viable option for deaf children—do not have to go together. Technology does not determine what is done with it: we do.

Finally, I explicitly collected the stories of mothers. This contributes to our understanding of the nuances of medicalization in contemporary life; it is also a feminist endeavor. Indeed, today families make choices about disability and biotechnologies in a contested environment. Mothering in particular is hard in this context. How do women live with the policies in place? How do current systems of implantation uphold the divides over the device and place increased demands on mothers? Investigating these and other questions is also a way to plan better; this inquiry has practical application for program evaluation, better-designed services, better policy making, and better meeting of people's needs. But the chapters also serve a reflexive purpose: they give us the opportunity to slow down clinical processes and interactions and to examine the consequences they have on people's lives.

1

A DIAGNOSIS OF DEAFNESS

How Mothers Experience Newborn Hearing Screening

The Other is—sometimes quite suddenly—
a member of the family.

• Gail Landsman, *Reconstructing Motherhood and Disability in the Age of "Perfect" Babies*

The routines of clinical staff and the lives of the families who participated in my study show that implantation is a long, emotionally draining, socially dynamic, and institutionally embedded process; that is, implantation involves a period of years that spans from the process of diagnosis to the long-term follow-up care needed years after surgery. Yet in popular culture, the CI is often seen as a miraculous and instantly effective surgical cure for deafness. Many of the most popular YouTube videos about the CI show it being turned on and someone "hearing for the first time." But implantation is not a onetime event, nor is the person in those videos experiencing the same kind of hearing that hearing people do. We also tend to assume that implantation occurs immediately after diagnosis of hearing loss. This is also not true; CIs are not a "first-line" treatment. Current standards of care require parents to try other methods of intervention first (such as hearing aids), and once a child is documented as having inadequate benefit from them, only then does insurance cover the CI and can candidacy for surgery be considered.[1] Even after surgery, however, a long-term commitment is required to learn how to use the CI.

Throughout implantation, the families in my study were socialized into the medicalized script of deafness as well as into performing various kinds of interventions. While there is no one CI story, every family experienced a point of entry into this pattern of socialization followed by movement through five similar stages: identification, initiation of intervention, candidacy, surgery, and long-term follow-up care. Some children were identified as deaf at birth and had this diagnosis confirmed within weeks, while others become ill, for example, and were diagnosed later. But regardless of how and when these children were diagnosed, it took time to socialize the families into the culture of implantation and to prepare the child and family for the device and for the years of habilitation after surgery.

Anticipatory Structures

Because interventions are multifaceted and long-term, the protocols and structures of the clinic remain consistent to facilitate the flow of parents from one stage to the next. I call these multiple strategies used by the clinic *anticipatory structures.* Anticipatory structures are persons, practices, and protocols in the clinic that are already in place and are triggered by a particular event and deployed to reduce parents' resistance to medical interventions. The main goal of anticipatory structures is to encourage and maintain compliance. They take many forms, like specific communication strategies used with parents, and staff cooperation with each other to provide different aspects of care to the same family. They have various purposes, such as ensuring follow-up visits and involving parents in the child's care. Finally, information sharing and knowledge transmission (whether from professionals to parents or from parent to parent) cannot be untangled from emotional support. For example, I observed clinical staff give medical—and at times highly scientific—advice while frequently describing what they were doing as emotional work. Clinic staff routinely anticipate the needs, concerns, and worries of parents throughout the implantation process. Thus, in addition to ensuring compliance, anticipatory structures also have a strong emotional component.

Compliance and self-regulatory behavior in patients, especially those with chronic conditions, are routine topics of health research. Here, parents assume the patient role on behalf of their child. But there is a complex relationship between the term *compliance,* indicating deference to medical power, and the term *adherence,* indicating that patients have agency or self-determination. For example, compliance is defined as "the extent to which a person's behavior (in terms of taking medications, following diets, or executing lifestyle changes) coincides with medical or health advice" (Haynes, Taylor, and Sackett 1979, 1). But this implies that patients are only passive recipients of medical advice who blindly obey expert suggestions. Critics think compliance suggests "that patients acquiesce to, yield to, or obey physicians' instructions; it implies conformity to medical or medically defined goals only" (Lutfey and Wishner 1999, 635). Proponents of adherence argue that it more accurately "captures the increasing complexity of medical care by characterizing patients as independent, intelligent, and autonomous people who take more active and voluntary roles in defining and pursuing goals for their medical treatment" (Lutfey and Wishner 1999, 635).

But the debates about compliance and adherence in both social and medical science also illustrate one of the central lines of inquiry in sociology: What degree of agency do individuals have as they live out their lives within social structures? The institutions that the families in this study navigate are an example of social structure; this shift to thinking about adherence mimics the shift to biomedicalization, where patients are seen as active participants in their own care. More specifically, how do families experience the process of diagnosis and how do they use the clinic as they frame the meaning of their child's deafness? In what ways do these institutions work upon these families to influence their utilization and framing?

In this chapter I delve into these questions by showing specific institutional processes, or anticipatory structures, that shape families' experiences. For the most part, families are unaware of the ways in which clinic staff strategize their communication styles and anticipate families' emotional response to the diagnosis. The following describes specific characteristics of anticipatory structures during the first stage of the process of implantation: identification.

Newborn Hearing Screening

The CI clinic at NYG is part of a larger hospital system, whose main hospital building is a short walk away. Both the audiology and otolaryngology (or ear, nose, and throat [ENT]) divisions are located in the CI center. The center's building is newly renovated, and everything inside is smooth and clean. The lobby is quiet and decorated with soft, pastel colors that soothe the eye. Just through this lobby is a set of glass doors that lead to the clinic's waiting room. The new, modern furniture continues the cool color palette, except in the kids' area in the waiting room. That area is bursting with bright colors and chock full of large plastic toys, wooden blocks, and tiny tables and chairs.

One morning, I walked through the lobby and down the hall to meet Margaret, the newborn hearing screening (NBHS) program coordinator. In every state in the United States, a NBHS program is in place to identify all newborns with a significant hearing loss. NBHS is a crucial part of the stage of identification, as failing a hearing screening is commonly the triggering event that activates the clinic's anticipatory structures for possible CI candidates. NBHS programs were implemented after a 1988 Commission on the Education of the Deaf report found that the average age of identification for profoundly deaf children in the United States was two and a half years old. At the time, hearing screenings, or tests that showed if a child had a hearing loss, were rarely used. As a result, children would often live for more than two years without any language input at all.

Pre-NBHS stories reveal an abundance of cases of late diagnosis. Lane, Hoffmeister, and Bahan noted in 1996 that the mother often sensed something amiss. For example, over the first year of a child's life, mothers "[might have] successfully and enjoyably play[ed] pattycake with their child, yet notice[d] that the child [did] not respond when urged to sing along" (Lane, Hoffmeister, and Bahan 1996, 32). Mothers might have repeatedly asked if something was wrong with their child, but their concerns might have been dismissed by a doctor, or they may have mistakenly been told that their child was developmentally disabled. Months would pass; "after repeated cycles of suspicion that a problem exists, rejection of the suspicion, and its re-emergence" (Lane et al. 1996, 33), finally the child would be taken for a hearing test.[2]

Today, identification practices are much different. After the 1988 report, the Joint Committee on Infant Hearing (JCIH) issued a position statement on the matter. The JCIH, composed of representatives from audiology, otolaryngology, pediatrics, and nursing, among others, recommended that children at a higher risk of hearing loss, such as those who were jaundiced or had needed neonatal intensive care, be screened before being discharged after birth. In 1990, the Surgeon General challenged state and federal agencies to devise a plan to have all deaf children identified before the age of twelve months. By 1993, the National Institutes of Health (NIH) recommended that all newborns be screened before leaving the hospital. In support of this, the JCIH wrote, "All infants with hearing loss should be identified before three months of age and receive intervention by six months of age" (JCIH 1994).

NBHS programs are now federally mandated in every state by the national Early Hearing Detection and Intervention (EHDI) program, which is housed under the Centers for Disease Control and Prevention (CDC). The federal EHDI program partners with state NBHS programs, which in turn partner with other state programs, such as Early Intervention (EI), a social services program created through legislation that provides therapy at home from birth to age three. NBHS programs' goal is to promote communication from birth for all children, starting with early identification of hearing loss. NBHS is only one example of newborn screening changes in recent years; numerous additional conditions besides hearing are now screened in newborns. Sociologists have already observed how the increased availability of screenings ensnares parents into a flurry of new kinds of medical decisions as they grapple with all kinds of new information (Timmermans and Buchbinder 2012). Today, more than 95 percent of all newborns in the United States undergo hearing screenings, and hearing loss is the most commonly detected "birth defect" (American Academy of Pediatrics).

As the NBHS coordinator, Margaret's role was to supervise the technicians who administer these screenings to newborns, and meet and follow up with parents when a baby failed the hearing test. That morning, we left her office and made our way to the main hospital building to visit the nurseries. We arrived at the maternity ward, where nurseries lined a long hallway. The staff called this area "Well Babies," the department where babies without any major complications or conditions were delivered. As we walked from the center to the main building, Margaret spoke of her work and how much she enjoyed it.[3] Clearly she was passionate about what she did, and she had been doing it for a long time. She suggested we go check on her "girls," meaning the NBHS technicians who make the rounds testing newborns.

We went through an unmarked door next to one of the nurseries and entered a small equipment room, where we found Cheryl, the NBHS technician, or "screener," on duty. She was gathering her equipment together on a two-tiered rolling cart. The metal cart was equipped with various machines and tools, stacks of pamphlets, and a pile of round, brightly colored "I had a hearing test" stickers. The stacks of pamphlets were separated into two groups: one kind was for children who passed the screening, the other for children who failed it. Before we began making our rounds with Cheryl, Margaret told me that because it was late morning, the babies would be with their mothers. This meant it might be hard to find a baby to test.

Margaret explained that there were different kinds of screening devices, and that the most commonly used one in the nursery is called an otoacoustic emissions (OAE) test. The small, gray OAE screening machine looked like a portable credit card machine, except for the wire coming out of it, which was attached to an ear probe with a rubber tip on the end. During an OAE, an earphone and microphone are placed in the baby's ear. A sound is presented; if the baby hears normally, the ear responds. This looks like an echo (or an "emission") reflected back into the ear canal, detected by the microphone, and measured by the device. If there is hearing loss, there is no emission. After the test, the machine produces a receipt displaying the results.

We were ready to roll out. Time was of the essence. NYG is a large hospital that handles thousands of births each year, and typically the screeners have only a forty-eight-hour window to work with: this is the amount of time insurance companies cover for well babies to be hospitalized. Margaret explained to me that early morning is the only time to screen the babies because they spend overnight in the nursery. After that, they are given to the moms. She also told me that if screeners miss that window, it is nearly impossible to get a screening in; having a baby is a celebratory event, and there are visitors and everyone is happy. If there is "bad news," she tells me she has to be very careful. But the upside, she says, is that EI is in place and ready, and it is NBHS that allows them to get children enrolled right away.

On a typical morning, the screener came in early and brought the babies to the testing room, because the tests require a quiet environment and a sleeping baby. If they passed, they would get a brochure

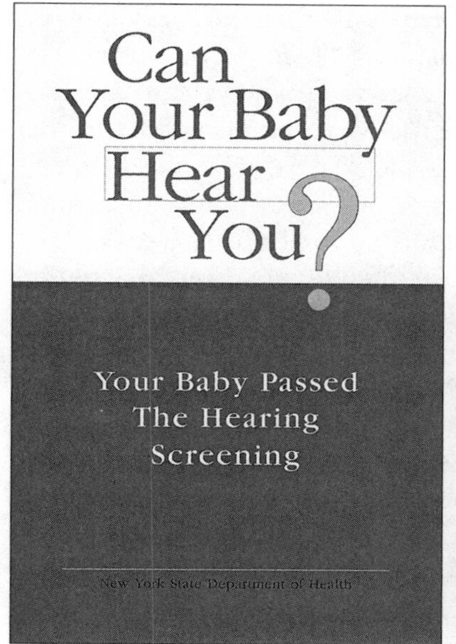

The front pages of two different pamphlets given to parents after the newborn hearing screening. Pamphlets published by the New York State Department of Health.

and be done, Margaret said, then added that if they failed the test, they would retest to make sure it was "a real fail." Because state law mandates NBHS, Margaret and her team of screeners are required to tell the parents the results. This is a highly managed moment. She advised that one should "never say fail," and the word *fail* does not appear in the pamphlet. Instead, it is framed as not being able to get a passing result yet.

Margaret emphasized further not to say the word *fail* to parents and instead to go in, sit down, and ask the parents if they could talk for a few minutes. This is when she would explain what it is the hearing screeners do and that they simply were not able to get a passing result yet. She also explained to me that she had developed ways to deliver this kind of news without making moms "hysterical." To do so, she would say that there were many normal reasons for not passing the test yet, such as fluid being in the ears if it was a Caesarean birth. However, Margaret warned that she had to be careful not to "blame" the C-section because then the mother might call her ob-gyn and complain. That could then get her in trouble with that doctor, and that was not something she wanted. She also told me that mothers always have the same questions for her; the first one always was, Is my baby deaf? I asked her what she did then, and she demonstrated to me in a most wonderfully soft, reassuring voice how she explained that it was not at all what they were saying, and that it was probably just fluid, so they should come back to the center in a month for a retest. She also assured parents that things are usually fine, but it is always best to follow up.

After we took a walk through Well Babies without finding any babies to test, we went to the neonatal intensive care unit (NICU). The NICU is another testing environment altogether. Margaret described it as a place for very sick babies. She explained that she had to know when a baby could tolerate the test safely, and she noted the importance of not antagonizing the nurses, who might have just spent an hour feeding a baby, who then throws up after being tested. At the NICU, there were babies in incubators with one-on-one nursing, some of them extraordinarily fragile. Many were on respirators and in large, enclosed pods with lots of wiring and tubing. Some of the machines were taller and wider than I. These babies are not brought into the testing room; a technician comes through with the cart. As we moved through the unit, we first rolled the cart to one of the last rooms, which holds the babies who are closer to being discharged. These babies are less fragile; there were about ten of them. The technician walked around, checking to see if it was a good time to test and whether the test had already been done. If it had, there would be one of those "I've had a hearing test" stickers on the top of the clear plastic container the baby was encased in.

Again, there were no babies requiring testing at that moment in this room, so we moved into the newer wing of the NICU. This section of the NICU is completely different, with its all-glass walls, state-of-the-art technology, and computers everywhere. Each of these babies had his or her own room, or "pod," and all were in incubators. Because all the walls were glass, I was able to look into all the pods. In one of them, doctors surrounded one of the incubators, performing a surgery. We rolled the cart from pod to pod; as we did

Cheryl checked charts to see if it was a good time to test. We came upon one baby that was in the NICU as a precaution because the mother had had a temperature during childbirth. We made our way to the incubator. Cheryl, Margaret, and I all peered into the clear plastic little box. He was sleeping, and he was tiny. Next to him was a nurse sitting at the computer. Cheryl powered on the equipment, took the wire from the OAE machine that had the ear probe on it, and placed it in the infant's ear for a few seconds. She then pressed a button on the machine; it lit up. A few seconds later it printed what indeed looked like a receipt: It read "passed" in black text at the bottom of the paper. Cheryl placed a sticker on the baby's name card at the top of the box. She then placed a "your baby passed the hearing test" pamphlet in the crib with him.

After this, Margaret and I left the NICU to return to her office at the CI center. When I asked her what happened if a NICU baby failed (as opposed to a "well baby"), Margaret said they tell the neonatologist but not the mothers, explaining that the mothers have a lot of bad news to deal with and that the neonatologist has a relationship with them that the screeners do not have. Margaret is clearly aware of and negotiating a lot of constituencies; she knows she has to manage the mothers and her own colleagues and to satisfy the law. She said that she had to put her head down and do her job, and that because the testing does not tell us anything about what kind or what degree of loss, her job is only to convey whether the baby has passed or failed.

When a child does fail a screening, Margaret makes sure that the mother brings the child back to the CI center for a follow-up test, usually within three months. To assist in accomplishing this, when a well baby fails the screening, she promptly shares this information with their pediatrician. She explained to me that pediatricians are far more effective than her team is at ensuring that mothers bring the children back for further evaluation, because they have established relationships with the mothers. This kind of careful relaying of information—anticipating emotions and managing mothers—is crucial to the clinical staff in general. Forging a relationship with "the moms" (a phrase used by Margaret and many others in the clinic) is the most important and crucial step in identification and the initial intervention: The possible need to engage in the long-term nature of the implantation process means connecting with families and building trust.

This is why, when a child fails the newborn screening, Margaret employs the communication techniques that she demonstrated to me, such as speaking softly and in a comforting manner, not using the word *fail,* reassuring mothers that the child could probably hear, but that they should just come back and confirm. But Margaret also makes sure that the mother has an actual appointment paper for that follow-up appointment in her hands before she is discharged from the hospital after giving birth. And the effort does not stop there. A team of secretaries assisting Margaret is devoted to making follow-up phone calls to these mothers, reminding them to bring their children back to the clinic. This is all an effort to stay on top of parents. She explained that they really keep after them, and once they get the moms in the clinic, they descend on them, and "we've got 'em."

Newborn hearing screening is the standard of care across the United States; however, national data from a 2007 study show that approximately half of those failing the screening are "lost to follow-up" (Russ et al. 2010). Although it may seem straightforward, the "system of care for infants and young children in which the program operates is surprisingly complex . . . experienced pediatric audiologists needed to perform diagnostic testing are in short supply; families, especially those in rural areas, frequently need to travel long distances to access definitive audiologic testing, which often requires several sessions" (Russ et al. 2010, S60). Thus, many families experience challenges in obtaining follow-up care, a pattern that is well documented (Honigfeld, Balch, and Gionet 2011; Yoshinaga-Itano 2003a; Van Cleave et al. 2012). This is the context in which Margaret is working. Other studies have shown high variation in the kinds of attempts used by clinics across the country to decrease the percentage of children lost to follow-up (Van Cleave et al. 2012). Some studies advocate for involving physicians in the process (Moeller, White, and Shisler 2006), while others focus on training parents through repetition of the message to return for follow-up, providing clear objectives and tasks (such as setting an appointment date), and using concise and simple informational materials like pamphlets (Honigfeld, Balch, and Gionet 2011).

The New York State (NYS) Department of Health, which administers EI, was awarded a CDC grant to enhance the NYS Newborn Hearing Screening Program's surveillance system in order to "decrease the number of children who are lost to follow-up in the newborn hearing screening process" (NYS Department of Health, Division of Family Health Bureau of Early Intervention 2013, 25). The stakeholders involved here are numerous, ranging from federal to state to hospital institutions, to health policy makers, down to the NBHS team Margaret manages. Margaret tries to anticipate all of the obstacles to parents' returning to the clinic so that she can try to resolve them. She also uses microinteractional adjustments and emotional labor to anticipate parents' reluctance and confusion.

Cultural Values behind the Practices

Rather than focus on the efficacy of the practices themselves, I want to highlight how these observations of the social organization of the clinic show that anticipatory structures operationalize larger cultural values and formalize a particular script about deafness; that is, the professional practices of the clinic draw from larger cultural narratives that assign negative meanings to deafness. The assumption that deafness is bad permeates Margaret's interactions with families as she uses strategies to manage the fear she anticipates families will feel and attempts to make sure that fear will not prevent them from following up. Pointing out that deafness is not inherently "bad news," however, *is not* an argument against identification or newborn screening practices. Rather, it is an acknowledgment that what identifying a child as deaf means is not fixed; different people assign different meanings to this diagnosis.

Avoiding the word *fail* and framing a follow-up visit as a way to get the passing result is particularly telling. It tells us that deafness is assumed by clinic staff to be negative and is immediately characterized as undesirable. Douglas Maynard (2003), a sociologist who specializes in conversational analysis, argues that relaying news is a highly social event that uses a variety of interactional techniques, including word choice and the suppression of facial expressions. In such interactions, meanings are socially constructed to produce "a mutual sense of some event-in-the-world as news and as having a good or bad character" (Maynard 2003, 27). And yet the meaning of deafness is not inherently bad. Many Deaf persons have expressed being thrilled when their child was identified as deaf. In 2002, the *Washington Post* featured a story about a Deaf lesbian couple who had purposefully used a genetically deaf donor to increase their chances of having a deaf child. "It would be nice to have a deaf child who is the same as us," one of the mothers states in the story (Mundy 2002).

However, the Deaf cultural script is not the dominant script in society or in the clinic. As the following section shows, the mothers of deaf children I observed draw from the same cultural scripts as the professionals. They participate in the same *latent script,* a script only enacted upon their children's failure of the NBHS testing, one that draws on presumptive, shared meanings of deafness. Margaret and other clinic staff are not the only persons constructing deafness as a medical problem; families share those same assumptions, so the script makes sense to all involved. Furthermore, these encounters are highly emotional; these hearing parents did not expect their children to be deaf, nor was deafness something for which they were prepared or to which they had prior exposure. Maynard also argues that bad news can be "interruptive (even disruptive) of the ordinary, taken-for-granted world . . . [as it] necessitates a realignment to and realization of a transfigured social world" (2003, 4). In interviews—as shown below—parents often described this moment as devastating, disorienting, and confusing.

This cultural and emotional work is sited within the institutional context. Even though health care institutions provide care, they are also places of work with their own demands, institutional culture, and politics. The NYG hospital system, for example, handles one of the highest numbers of births annually in New York State. Margaret typically has forty-eight hours to test a newborn, and she also has to manage the turf and feelings of her colleagues (such as the nurses in NICU and the pediatricians). She also has to manage the emotional reactions of mothers and families, find ways to communicate failed tests, *and* ensure that mothers follow up with the audiologist. While performing emotional labor work, she also has to ensure the smooth machinery of the institution. This is all necessary for her to be able to "keep her head down" and "do her job."

The social technologies deployed in this aspect of the identification phase are centered on efficiency, compliance, and anticipation; Margaret and her team of screeners and secretaries deliberately stylize the giving of "bad news," undertaking a tremendous amount of anticipatory emotional work. Expecting parental grief, denial, and/or refusal of follow-up testing are all important aspects of Margaret's task to ensure compliance. Similar anticipatory structures also continue in the next step.

Finding Out: The Follow-Up Appointment

Sometimes the follow-up appointment confirms that a child has hearing loss, but it is actually much more common to find that the child *does not* have hearing loss. The figure varies annually and from state to state, but approximately only 25 percent of children who fail the newborn hearing screening go on to be identified with a hearing loss. If there is a hearing loss, the follow-up appointment is often where the loss is confirmed.

When parents bring the child back in to retest, they mostly likely come to the audiology clinic for an auditory brainstem response (ABR) test. The best description of the ABR comes from Lisa, one of the audiologists in the clinic whom I observed during appointments. "ABRs are like EEGs or EKGs; they are just measuring electrical energy that moves from the cochlea to the brain stem—that's why it's called an auditory brainstem response test. It's just like when you get one done on your heart and the stickies go on your chest. With the auditory system, the stickies go up around your ears."

ABR tests are typically performed at the clinic follow-up appointment after the child has failed the NBHS. However, a child might be diagnosed later through an ABR as well, for example, after another precipitating event like a postnatal high fever or illness, or when a mother suspects a hearing loss. A failed newborn screening is not the *only* precursor to an ABR, but it often is one. I sat in on ABRs to understand how mothers and audiologists interact during these appointments. Sharon, the center's director, provided me with information for my first observation of an ABR test. This one was an NBHS failure from a nearby hospital for a baby born via Caesarean section. She told me that there was some question as to whether it

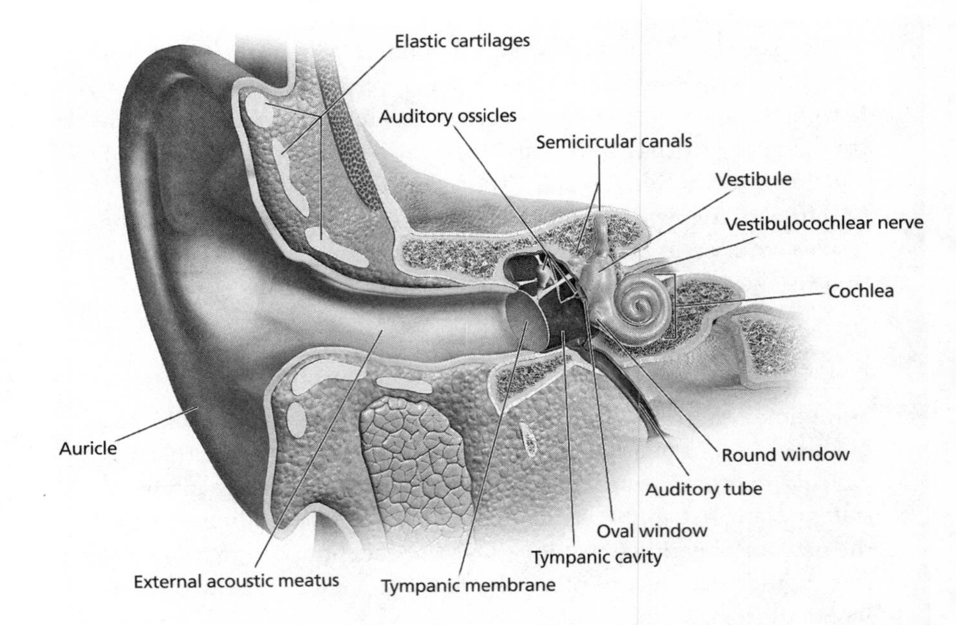

The Anatomy of the Ear

A cross-section of the anatomy of the ear, including the cochlea and auditory nerve. Illustration by Bruce Blaus, Blausen Gallery, 2014.

was a true failure. She told me that we were starting from scratch with this particular baby; all that was known at this point was that the child had failed the screening.

Sharon warned me that the tests could be difficult. "Often the mother will hold the baby [during the test], sometimes they do better in the car seat if they are sleeping, so we just leave them in that, or if the mother wants me to hang the baby from the chandelier . . . I do whatever they want." I quickly learned that the reason for this is that the child has to be in a natural sleep state for this test—which can take a long time. A natural sleep state is, of course, not always easily attainable in infants, making this test even more emotionally stressful for parents. Meanwhile, audiologists sometimes seem especially annoyed by ABRs; the time they take sometimes backs up appointments, potentially disrupting work flow. Thus, before ABR appointments begin, tensions are typically already rising for both audiologists and parents due to the nature of the test, its length, and the possibility that it could take hours to get a proper result.

I accompanied Sharon to the waiting room, where a woman sat with a pink stroller and her husband. She looked tired. After Sharon explained that I was conducting a study and would be observing the appointment and got her permission for me to be in the room, she explained that only one of the parents could come with us. They conferred and decided that the mother should go with us. The three of us walked the stroller down the long hallway to a room labeled "Electrophysiology." The door is about eight inches thick, and the room looks like a giant vault. The entire room is soundproofed; the walls, floor, and ceiling are covered with carpet. There is a recliner for Mom, a table with a computer screen on it, a long countertop with various machines and computers, and a small chair by the door for me. The first thing Sharon explained was that nothing she did in this test would hurt the baby. She then put electrodes around the baby's head and told the mother that the gel she was using to attach them was cold and was probably the most uncomfortable part of the process. When I sat in on other ABRs with two different audiologists, they also calmed mothers right away by assuring them that the test did not hurt. As Sharon attached electrodes, the mother said that she believed her daughter could hear because she seemed to respond to loud sounds. Sharon did not respond to this comment, staying focused on her task. She started with the right ear since that was the one that initially failed the NBHS. She asked me to turn the lights out. It was eerily quiet in this soundproof room, illuminated then only by the computer screen. Sharon sat down at the computer, manipulating the data that appeared in the form of moving and waving graphs and spectrograms. As though anticipating the mother's questions, Sharon said she would not be able to tell her the results during the test. The audiologists at the other ABRs I attended also stated this up front. One audiologist immediately told the mother that she would not be able to tell her anything from the results until the very end. As I watched Sharon conduct the ABR, no one said a word; the only sound was the clicking of the mouse as she arranged and rearranged the squiggling lines on the screen. I was sitting just several inches away from a tiny girl who was only a few weeks old. She had electrodes on her head and was breathing audibly, sleeping.

The mother looked exhausted, but we were all relieved that the child was sleeping peacefully. Over the course of the test, when the baby cried out a little or moved about, Sharon reached over and gently rocked the stroller for a few moments until the infant fell back asleep. The mother peered steadily at the computer screen, as if to try and decipher all the different colored lines. The anticipation was palpable. Sharon stayed focused, clicking away, tagging points on the graphs and separating the lines.

About thirty minutes later, it was time to switch to the other ear. Sharon removed the miniature earphone from the baby's ear and put it in the other. This time she connected the computer to the electrode attached to the other side of her head. To do this, she had to turn the child's head, and she very softly rocked her. The mother looked on, letting Sharon do the soothing at the moment. Another half an hour later, after the second ear was finished, Sharon moved to another machine in the room to conduct tympanometry, which tests the middle ear, specifically the functioning of the eardrum, or tympanic membrane.

When this was over, Sharon said she wanted to unhook her first from the electrodes so that Dad did not have to see her "looking like Frankenstein." She went to get the father, and I sat alone with the mother. She smiled at me, told me she was exhausted, but that she thought the baby did really well. A moment later, the dad came into the room with Sharon and sat down. He anxiously asked how it was. "Good news: The baby hears," Sharon announced. She told them she was picking up a mild hearing loss in both ears, but it was only slightly shy of where "she should be normally." They proceeded to talk about the possibility of removing fluid from the ears to solve the problem, and about seeing Dr. Brown, the ENT doctor at the center (who is also the surgeon) at the clinic as soon as possible. It was after this appointment that I learned that audiologists are not medical doctors but allied health professionals who cannot make a diagnosis. Thus, all ABRs that indicate hearing loss must be followed up by a trip to the ENT for a diagnosis.

Before they left, Sharon asked them if there was anyone in their family with a hearing loss. The mother said no and explained that they were puzzled by the possibility of their daughter being deaf. The father worried that it was a permanent hearing loss. Sharon explained that the clinic recommended hearing aids and that they needed to address the hearing loss early, and she also assured them that we were "getting ahead of ourselves here." She gave them Dr. Brown's card and told them that no matter the outcome with ENT, to come back and see her. As Sharon stepped out, the mother looked at me, wide-eyed, and said how worried she had been that her daughter was deaf and how relieved she was. I could see the relief on her face and the tears in her eyes. She turned to her husband, and he put his arm around her, and they began to walk down the hallway.

The ABR tests can most easily be summed up as connecting wires, putting probes in the ear, pressing buttons on the computer, waiting, and repeating. But the tests do not always go smoothly. Sometimes the child wakes up and will not go back to sleep. When this happened, I saw parents—already exhausted from having a newborn and the stress of the test—worn further and further down. I also witnessed audiologists grow increasingly frustrated as appointments got pushed back and the waiting room backed up. ABRs, already tense, are layered with competing tensions between the parents' and audiologists' needs.

In the cases of ABR appointments where a child was crying and could not sleep, the audiologist left the room, headed to the shared audiologists' office, and gave parents time and space to get the child back to sleep. These moments back in the office can be lively, as the audiologists' offices are nestled together in a corner of the center. They gathered to express frustration and vent. They made phone calls, filled out paperwork, and dealt with insurance issues for other patients. Sometimes they discussed giving up on the child and rescheduling the ABR, or if it was an older child, having it done with sedation. An anesthesiologist has to be scheduled to attend an appointment with sedation. Sedation does pose a risk for the child, but in the interest of convenience for the clinic's daily operations, this is generally not seen as prohibitive. Audiologists' concerns are efficiency, completion of the test, and multitasking to serve other patients and handle a busy appointment schedule—all of this while simultaneously parents are on the other side of the vaulted door, experiencing emotional duress.

In an ABR that I observed with Holly, another audiologist, the child did not cooperate, so we left the room to give the mother a chance to get her to sleep. As we walked down the hallway to wait, Holly told me that when the OAE was being done, I should look for blue. Blue, she told me, was a good thing and indicated normal hearing. I would see red, but then I should look just above that for the blue. Here was an odd moment for me as an ethnographer: I was now privy to information about the child before the parent would be. As I sat with Holly in the shared audiology office, I felt complicit with her. A few minutes later, Holly decided it was time to check back and see if the baby was sleeping.

Upon entering the ABR chamber, we found that the baby still was not sleeping. The mother was quiet. Holly did not explain much, except that she was going to go ahead with the OAE. I watched the machine. I did not see any blue, but I was also not quite sure where I was supposed to be looking. A few moments later, Holly asked me to press the buttons on the machine as she sat across the room with the child, manipulating the earphones. She said she was going to put me to work and asked me to press a button. I pressed the start button on the tympanometer. A few moments later, I did this again for the other ear. Once again, I felt strange; on the one hand, I felt pleased to have some kind of function, since I suspected that, as the ethnographer, I was seen to be taking up space and possibly annoying the clinic's staff. On the other, I felt conflicted at the thought of participating in clinical routines that had the goal of determining whether a child was "normal" or not.

When I observed an ABR with Lisa, the mother that day was more overtly anxious. She hovered over the screen and made suggestions as to how to make the baby more comfortable. She offered to hold equipment for the audiologist, suggested how to position the baby, and wondered out loud whether they should start with a different kind of test. I noticed that Lisa simply did not respond to most of her comments (a technique I saw often employed in all kinds of audiology appointments). Once Lisa got the ABR going, the mother fixed her eyes on the screen and kept asking whether her child could hear. Despite Lisa having told the mother that she would not give results during the test, the mother asked anyway; Lisa did not respond.

Finally, at the end of the test, Lisa announced the results. She said that she still could not get him to pass the screening. She also told the mother that the test was indicating loss in the high frequencies in the left ear, but that the right ear was responding better than last time. However, it still appeared to show a "severe to profound hearing loss." She told the mother that she would need to make sure and do more tests. The mother seemed confused and did not respond. Lisa continued that it was "always best" to go through life with two normal hearing ears. But the good news was that only one normally hearing ear was enough to develop speech and language. This was a tricky moment for many reasons. On the one hand, the results showed a severe to profound hearing loss, possibly in only one ear. But the results were still generally framed as simply not being able to get a passing result yet. Lisa gave the mother her e-mail address and said she would speak with the ENT and could be in touch with a plan later that afternoon. Then she left the room. I watched as Lisa walked down the hallway, and the mother, in a kind of daze, packed up her belongings and readied herself to leave. I was not sure what she was feeling in that moment. Did she feel hopeful since Lisa had said there was good news that one ear was not lost? Did she feel that the passing result would eventually happen?

Mothers' Experiences of the ABR

Since the parents in my study were those who eventually got a CI for their child, I wanted to know the answers to the question, What did they feel when they got the diagnosis in that ABR room? I visited Carol, a white woman in her forties who is a professor at a university near her apartment in Queens. When I arrived, her hands were full of sewing materials—fabric, thread, and so on. We walked into her kitchen, where she set the sewing items down on the counter while she prepared some tea. Then she asked if we could move to the living room so she could finish the project she was working on. She plopped down on the couch, picked up a piece of clothing from a pile of baby clothes next to her, and grabbed her needle and thread. On her other side was a pile of baby socks. She started to sew one of the socks into the top of the baby outfit. "The implants have external controllers," she explained. "Your unit has the microphone and that goes on his ear, but because it's a body-worn controller, because babies ears are smaller," she said, "it sometimes falls off." So, sewing socks into his baby clothes created a place to keep the external piece.

"It just slips right in," she said, quite pleased with a big grin across her face. "Plus, because [the sock] is sewn into his shirt, then when he does pull it off his head, we're not going to lose it. So it just makes life a little bit easier."

Carol recounted her NBHS and ABR story. Her son, Jeremy, had been screened at another local hospital, not NYG. When the staff said that Jeremy had not passed the test, she asked what that meant. The staff's response was, "He probably has fluid in his ears. You know, it was a C-section." Jeremy was tested three times. "By the time we're leaving [the first time], we're like, 'Hmm, so he hasn't passed that hearing test,'" Carol said. The screener told her, "It's fluid, whatever, just make sure you go back." They went back. Jeremy failed again a month later; the screener told Carol she needed to wait a little longer, perhaps there was still fluid. They went back again. He failed again. "So we made another appointment [at NYG], and, honestly, it never occurred to me, even though he kept failing, that—because you know what I thought they were going to tell us was 'He has to have some sort of procedure to clean out the fluid,' or something like that." On the morning of the ABR appointment, it was the NYG director, Sharon, who conducted Jeremy's test. Carol said that as she walked into the center, "It struck me. I thought, hmm, well, maybe this is something."

In the ABR room, Jeremy slept on her lap. "I'm thinking, he's not really making so many movements here. I'm wondering [about his hearing], but I don't know," Carol related. They sat through the ABR and then, "The poor woman, Sharon, says—and I say poor woman, because it's kind of a funny story. She says to me after it's over, 'Jeremy has a severe to profound hearing loss.' Now, understand that I know what that means now, but that didn't mean anything to me then, really," Carol said. In fact, Carol thought that such a loss might be reversible, since she was still under the impression that it was simply caused by a buildup of fluid. "I understand what deaf means, but I don't know what 'severe to profound hearing loss' means in that moment." Plus, Carol tells me, Sharon was whispering. "I'm thinking, I think that means he's deaf, but it can't mean deaf, because she's whispering. You know, that's what's going through my mind. I'm thinking she's contradicting herself, because I think she's saying that he can't hear anything. But then why would she be whispering?"

Carol continued, "I said, so you mean he's deaf? And she says yes. But [I ask] then why are you whispering?" Carol went on to relate that she suspected Sharon thought she was some kind of "wiseass" for making that kind of remark, but that that was truly what was running through her mind. Right after that moment, though, "I just got this kind of gasp to myself, because one thought went through my mind: Oh my God, other little kids won't want to play with my baby. It's going to be so hard for him," Carol said. All of my parent participants related this same terror of the future and fear of lack of acceptance of their child. Multiple times parents conveyed to me their vision of an isolated future for a deaf child and similar sentiments.

To combat this fear, she said she thought to herself right then: "I'll teach him how to play baseball really well, then other kids will play with him. . . . We can teach him how to be a really good athlete or be good at something, and then that problem's solved. So I thought, OK, I'm good." Then Sharon started talking about CIs, right there, after the diagnosis. Carol said that she just nodded and said "OK" and "all right" to everything she said. Even hearing about implants, however, did not mean anything to her in that moment. "I'm thinking, OK, I'm going to have to talk to you about this thing you're talking about. I'm going to have to learn sign language. I'm going to have to do these things. Then I think, oh my God, I've got to tell my husband, who is sitting in the other room."

At this point, Sharon went to get Carol's husband. He came back to the testing room and became very upset. Carol remained calm and told him to go and make sure his parents, who were also in the waiting room, were also calm. Then, immediately, they all together met with the center's social worker, Sonya, in Sonya's office, down the hallway from the ABR room. Carol remembered, "She starts telling us about the CI and that he might be a good candidate, and we have this service and that service." In fact, she felt Sonya was "incredibly helpful, she was really great, very knowledgeable, very calming. She gave us a lot of information, saying we're here for you; she was very supportive." Carol knew that she needed to listen carefully to all the information so that she could perform the impending tasks. She also, however, emphasized to me how thankful she was that Jeremy was "just deaf." She had concerns that because she was forty when she had Jeremy, there could be something "wrong" with him. Carol said that she thought to herself upon Jeremy's diagnosis, "You're telling me he's deaf, but there's nothing else? That you think that's what's wrong? So, OK, he'll be deaf. You know, some people are left-handed, that'll just be that. And the cochlear implant thing, I had no idea what she was talking about. I'm thinking sign language."

In ten interviews with parents whose child had confirmed hearing loss and went on to get a CI, every parent described the end of the ABR as the moment their child's hearing loss was first officially communicated to them. The most common characterization of this moment is devastation and grief (although Carol's account may be mitigated by her belief that she was taking a risk by having a child at forty). Jane, Lucy's mother, told me about Lucy's diagnosis in an ABR with Sharon. "I went to pick up Lucy—and she was sedated—our eyes met, and Sharon said, 'She's deaf.'" Jane continued describing it as "one of those flash moments with me in my life where I just—I remember making eye contact with Sharon . . . I just fell apart. I remember the wave of grief." Another mother, named Becky, described how an audiologist diagnosed her child as deaf after an ABR. They had had to go back to the clinic and attempt the test three times because her daughter Amy would not sleep during the test, which had already increased Becky's stress. Then, Becky said, "They said that she was profoundly deaf in both ears . . . it was a big blow." Like Carol, Becky was immediately given information about the CI. "I really didn't know at the time [of the ABR] about CIs. I had no idea." The audiologists immediately explained the process of implantation in full detail at the end of the ABR session: "They explained everything . . . and they told us that the criteria was that Amy had to go through the hearing aid process first—which didn't do anything for her—they told us you have to go through these steps first to see if she qualifies. They gave us the number and the address and stuff for the implant surgeon. They really wanted us to see the surgeon."

Anticipatory Structures in Identification Practices

The end of the ABR test often offered a definitive answer as to whether the child was deaf; any confirmation of hearing loss in an ABR was a crucial triggering event for future anticipatory structures (which are explored in the next chapter). This is exactly why the clinic makes such efforts to ensure follow-up as soon as possible after the NBHS and to take care of parents' emotional needs during diagnosis. During this stage, professionals use various techniques, such as careful communication, future appointment dates, follow-up calls, emotional support through a social worker, and, during diagnosis, the dissemination of highly technological information about interventions like the CI.

All clinic staff acknowledged that parents are especially vulnerable at the diagnosis stage. Many are sent into a tailspin of grief, and as a result, the backstage work that organizes parents' experiences from this moment forward becomes more complex. This reaction to a diagnosis of deafness has been well documented elsewhere (e.g., Fjord 2001). As a result, the reach of anticipatory structures from this point onward shifts far beyond just getting parents back to the clinic for a follow-up. The structures begin to web outward, including disseminating information, instilling new forms of parental competence, and providing various sources of emotional support. To execute these structures, parents are asked to schedule the next follow-up with the ENT doctor, as audiologists actually cannot make an official diagnosis. Thus, the protocol is to ensure that parents leave the final ABR appointment having scheduled a future appointment with Dr. Brown, the clinic's ENT doctor and CI surgeon. Parents are also sent directly to speak with Sonya, the social worker, or encouraged to contact her if she is not available. As will be seen in the following chapter, her role becomes much more central, as she connects parents to the New York State EI program. This connection with EI then creates the cycle: the expectation of future appointments, the providing of emotional support, and the perpetuation of the mother's relationship with the clinic. As Margaret indicated earlier, the explicit goal of the clinic staff is to have the mothers come back to the clinic and to envelop them in the expectations of the medical script of deafness, which may include CIs. Finally, in an effort to combat the panic that parents may experience upon hearing that their child is deaf, clinic staff may immediately provide a deluge of information on implantation.

All of the parents in the study spoke to me about how they were given a lot of information about the CI at the end of the ABR or very soon after. The way parents put it, they were given a "path": a plan or a series of steps to begin working through. It could mean that staff believe they can mitigate that devastation by conveying how deafness is combatable through future medical intervention. But based on my conversations with audiologists, they also do this so that parents can get started right away on completing all of the necessary steps before the CI is placed—in other words, to comply with the path set out before them. It is about orienting the parents to their new "reality," as the chief audiologist put it to me. Like the "bad news" that Margaret sometimes has to deliver, this unequivocally sends the message that deafness should be corrected in a specific way. It also suggests that if you do the work, it will be successful.

While anticipatory structures are present throughout this stage of identification, they also precede it. This chapter focuses on clinic-specific anticipatory structures, but the broader NBHS protocols that were legislated in the 1990s can also be seen as broader anticipatory structures preceding and framing these clinical events. Programs like NBHS, although they operate on a much larger scale, have also been framed by scholars as biopolitical (e.g., Lemke 2011; Foucault 2010; Rose 2006), wherein science and scientific knowledge merge with normative ideas; that is, science and medicine are enacted on a population level based on concepts that invoke our shared, latent script on what normalcy is. This is illustrated here by how concepts such as deafness and disability are categorized, responded to, and institutionalized in formal, systematic ways by the state—and thus in the clinic.

Foucault argued that biopolitical regulation is achieved by framing certain conditions as health issues. Once the state deems something a health problem, then it becomes reasonable and rational to act upon it. As a result, citizens tend to assume that governments are responding to problems and/or preventing them from happening in the first place by focusing on early identification and/or anticipation. During this process, the problem is created precisely through how it is named and framed discursively, for example, in a medical framework and with medical language (Foucault 2010). Although Foucault and other scholars interested in biopolitics may not use the term *medicalization,* medicalization is clearly an aspect of biopolitics. Rather than examining this from a macroscale—such as population-level theorizing about biopolitics—this chapter focuses on the microscale by showing how families live with preset policies and how they interact with institutional-level anticipatory structures in a clinic.

Latent and Neurological Scripts

It is frightening for families to imagine the future for their deaf child (a theme that recurs throughout the following chapters), and especially frightening to think about not being able to communicate with their child through spoken language like hearing parents of hearing babies do. Professionals anticipate this fear. Because parents and professionals are both part of a larger society that has dominant scripts about deafness and disability, both parties hold similar meanings around deafness. Thus, throughout the identification phase, parents and professionals both draw on similar latent scripts regarding deafness—scripts that do not include a knowledge of the Deaf community and the history and culture of sign language, and perhaps interest in either of those things. Unless parents of deaf children are also Deaf, they most likely encounter deafness for the first time through the medicalized script that is currently formalized in our institutions.

Kathryn Meadow-Orlans and colleagues (2004) found that in the past, when deafness was often not diagnosed until much later, hearing parents had spent time with their child and had already adapted visually without necessarily naming the deafness for what it was. Fjord found that after diagnosis, however, parents experienced a loss of competence, as she explains: "Before a diagnosis of deafness, hearing parents had already, and without knowing they were doing so, adapted to the visual needs of their child. . . . they were visually engaging their child, using gesture and touch that brought satisfying communication within the parent-child dyad" (2001, 133). When identification of deafness occurs much earlier in the child's life, the meaning of deafness is not only made in a medical environment but also in new ways than before.

Today, this space of adapting visually is often eliminated because of the changes in legislation regarding NBHS and clinical practices that anticipate deafness from a much earlier point. As a result of earlier diagnosis, the opportunity to experience a "loss of competence" by mothers of deaf children is no longer available; instead of losing one kind of confidence, these mothers are early on systematically instilled with a competence that reflects scientific motherhood. With implantation floated as a future goal from the start, mothers are offered a whole new scientific language and set of practices centered on developing spoken language as well. The tensions between these types of knowledges—the voice of medicine/scientific knowledge versus mothers' knowledge of their children—continue to play out in the stages that follow. The consequences of new techniques and protocols for hearing screening are thus ambivalent; we can now know if a child is deaf much earlier, and this is a positive change, but a particular script of deafness also supplants mothers' previous experience of visually adapting.

Furthermore, not only is the meaning of deafness made through a medicalized script, but it is also coconstructed through the technology of the CI in a neurological script. Starting with the ABR, the brain is at the center of diagnosis. As will become clear over the following chapters, neurological discourse is present from the beginning and continues throughout the stages of implantation. Parents begin their understanding of deafness as a condition of the brain, and discussion of CI follows; the CI is billed as the interface through which to correct the frequencies that have gone astray or are missing. Thus, it becomes a programming "patch" of sorts. If the ABR indicates a missing signal, the equipment (the CI) is implicated; it is ready and waiting to be installed to provide access to the brain in order to supply the missing signals. As I argue throughout the rest of this book, what accompanies this understanding of deafness is the idea that the brain must be molded in an explicitly *nonvisual* way.

Understanding Ambivalent Medicalization in Action

There are no disagreements over whether a child's deafness should be identified as soon as possible, but as I outlined in the introduction, there are different perspectives regarding the meaning of deafness once identification occurs. The main idea behind the Deaf cultural script of deafness is that deafness is a language difference (a social problem) rather than a medical problem that needs medical intervention. Deaf studies scholars and Deaf activists have developed a term to encapsulate the history of society's systematic response to deafness: *audism*. According to the website of Audism Free America, *audism* is defined as "attitudes and practices based on the assumption that behaving in the ways of those who speak and hear is desired and best. It produces a system of privilege, thus resulting in stigma, bias, discrimination, and prejudice—in overt or covert ways—against Deaf culture, American Sign Language, and Deaf people of all walks of life." In this view, identification practices that lead parents to intervention strategies that include the CI depend on "a particular cultural orientation favoring its own sensory orientation in the world" (Valente, Bahan, and Bauman 2011, 248). In one of two reports issued by The Hastings Center, a bioethics think tank, each taking an opposing view on the bioethical debates over medical intervention with deaf children and CIs, Crouch (1997) echoes the Deaf community's calls to embrace diversity and cautions readers not to be overly confident in medical intervention. This perspective sees current identification practices as fundamentally flawed because medical professionals generally adhere to the latent medicalized script that deafness is bad, use negative terms and medical language to refer to deafness, and do not routinely refer families to Deaf adults, Deaf community resources, or ASL resources.

But the opposing view that deafness is a medical problem is what drives the identification and intervention that commences in the clinic. The countering essay issued by The Hastings Center argues that early identification and implantation provides an invaluable opportunity to correct hearing loss via available technological means but also emphasizes identification and implantation as acts of social and parental responsibility (Tucker 1998). If parents do not opt for implantation and normalization through emphasis on learning speech, Tucker argues, then they should not receive support services and accommodations. The reasoning here is that opting for sign language leans on the state for interpreting and educational costs. Not giving children a CI is constructed as limiting their opportunities and depriving them of a basic sensory capability. Thus, the medicalized script of deafness sees it as straightforward that deafness is a problem of the child's body and a condition that should be mitigated through medical intervention as quickly as possible.

One problem with the claims of both of these scripts is that they are both too narrow. Ambivalent medicalization recognizes the capabilities of medicine and the relief it provides, while also acknowledging the consequences of this path. For example, hearing parents experience grief over their child's deafness; the diagnosis is often unexpected, unfamiliar, outside the realm of their experience, and requiring of a daunting skill set they simply do not have (for example, most hearing people do not know sign language). As a result, parents routinely express fear of deafness and of it being something "wrong" or different about their child. Thus, in these clinical encounters, parents find it extremely frightening to have it suggested that their child is deaf but also soothing to know that there is a set of actions that they can take to address the deafness. But the medicalized script of deafness suggests actions that are limited to medical intervention; these actions exclude the vast knowledge, resources, and experience of Deaf people themselves. So while families experience the benefits of medical intervention—such as support services and emotional relief—the boundaries of how they can frame deafness are highly monitored.

The medical viewpoint—with its clinical thinking, scientific language, and demand for particular types of ongoing, therapeutic labor designed to "overcome" the deafness—*displaces* other possibilities and ways of seeing. In the following chapters, we will see how this medicalized script of deafness is turned into a broader and ongoing way of life, as the demand to engage in scientific motherhood permeates the task of raising deaf children. Scientific motherhood in this context requires waiting for the next step of the process, having faith that the ongoing and uncertain medical interventions will pay off in the future, and often dealing with overwhelming amounts of care and technological know-how.

2

EARLY INTERVENTION

Turning Parents into Trainers

Well, if I'm lying on the floor having a panic attack, I'm not exactly making a healthy dinner.

• Jane, mother of Lucy

The maintenance and promotion of personal, childhood, and familial health—regimen, personal hygiene, healthy child-rearing, the identification and treatment of illness—are central to forms of self-management that authorities seek to inculcate into citizens and hence appeal to their own hopes, fears, and anxieties.

• Nikolas Rose, *The Politics of Life Itself*

During intervention, the next stage of implantation, anticipatory structures extend outward from the clinic; as a result, they create an ongoing therapeutic culture. In this chapter, I show the organization and ethos of this therapeutic culture at NYG, especially as it centers on the New York State Early Intervention Program (EI). Interinstitutional cooperation between clinic, state, schools, organizations, and the parent community all act as backstage efforts to support parents. Although these groups execute anticipatory structures because of professional and institutional pressures regarding efficacy and compliance, they also act from a sense of the paramount importance of tending to parents' emotional needs. Thus, this chapter also focuses on some of these "soft" aspects, such as how parents depend on other parents for emotional resources and support. Looking at the particular therapeutic culture at NYG, which is shared by professionals and parents, allows for understanding the larger structures organizing this specific clinic, as well as how the experiences of those participating in it are situated within a larger social context.

Early Intervention and the "Therapeutic Mode"

At NYG, it is standard practice that diagnosis be followed up with an immediate consultation with Sonya, the social worker, and she is involved in a wide array of services from that point forward. She is a therapist to the parents, facilitates their connection to other parents through support groups, and is generally available to them to tend to all aspects of their emotional process over the long-term. Clinic staff consider parents' emotional well-being of utmost concern; they discuss daily the ongoing need for parent counseling. And Sonya is an incredible resource; clinic staff and parents all regularly praise her tremendous skills at being able to offer parents both information and comfort. She is the nexus that links the highly structured intervention services in EI with the culture of emotional support for parents. When I sat down with her, I could understand why she was so good at this role. She is a wonderfully organized person—no family would "slip through the cracks" on her watch, she would say. She has been doing her job for many years, and she has the most soothing voice I have ever heard. She is warm and friendly and makes a tremendous effort to help the people around her feel comfortable.

To address parents' emotional needs, Sonya meets with them individually as soon as possible (sometimes moments after diagnosis, as with Carol); she refers them to the center's parent support group, or even elsewhere if that seems better for them. But Sonya also works as a Service Coordinator for EI, which is administered by the New York State Department of Health and provides home-based services to children under three years of age who have a confirmed disability or delay. Regardless of the etiology of the hearing loss, families with a child who has a qualifying hearing loss are eligible for EI services.

These intervention services are highly structured and highly monitored. According to the latest national numbers from a CDC study, in 2012, 85 percent of children identified with a hearing loss were referred to EI (CDC 2014). The implementation of NBHS programs since the 1990s has dramatically increased enrollments in EI programs, and, in turn, enrollment in EI services is likely one of the contributing factors leading to a greater number of children receiving CIs. Although this must be evaluated state by state, one study of the impact of an EI program in Kansas notes: "The percentage of the [EI] caseload with profound hearing loss who received cochlear implants more than doubled from 1998–1999 (42%) to 2005–2006 (100%)" (Halpin et al. 2010).[1]

At its core, EI claims to be a "parent-centered program." EI is not only about providing speech or various therapies to the infants enrolled but also about assisting parents in a variety of capacities, such as working to resolve access to transportation, food, child care, and additional medical services. EI is also training intensive. Speech and occupational therapists offer in-home "parent training" on how to practice daily therapeutic tasks with their child. Sonya explained that service providers and therapists—such as the ones contracted through Strivright, the Auditory Oral School of New York—go into the home and train parents "in everything." She continued, "In other words, you can't think that if you bring your child an hour a week for therapy that that's going to be the fix." In these trainings, parents are given a chart with goals and daily exercises, such as testing their children's response to their name or their ability to localize sound (this kind of work is dubbed "auditory training"). Sonya noted: "So we're saying, you know, we need you to be an involved person. It is important to keep at this, you need to be on it." However, she continued, "for whatever reason, [if] a parent doesn't want to come back, I'll try to help them. Maybe it's a transportation problem, maybe it's an insurance problem, maybe they're just being resistant."

When parents participate in auditory training, they are asked to create quarterly parent progress reports documenting their activities in accordance with the training plan. Parents return completed forms to their EI service coordinator (Sonya), who in turn provides the clinic with information to monitor the progress being made at home. The guiding philosophy of these EI services for the deaf children is that they should receive speech therapy and auditory training. The goal is to train parents, through a therapist, to intervene in a therapeutic and scientific way that reflects the medical script of deafness, and to do so on their own. Strivright, the Auditory Oral School of New York, is one vendor that contracts with New York State EI services; Strivright representatives work directly with Sonya and are

NYC EARLY INTERVENTION PROGRAM
FAMILIES AS PARTNERS (FAP) CALENDAR

Name of Interventionist/Agency	Authorized Service

CHILD'S NAME: ______________________ (Last) ______________________ (First)

EI #: ____________ DATES: FROM ___/___/___ TO: ___/___/___

FAMILY PLAN Month of ________ Completed by Interventionist(s). Number the activities.	Questions about Family Plan: What worked well in the plan? What didn't work? Comments, concerns and adjustments. (Completed by Parent/Caregiver)	Parent/Caregiver: List the number of the activity you tried. Put "+" if the activity worked well and "–" if it didn't work well. (Completed by Parent/Caregiver)						
		Sun: week of	Mon	Tues	Wed	Thurs	Fri	Sat
		Sun: week of	Mon	Tues	Wed	Thurs	Fri	Sat
		Sun: week of	Mon	Tues	Wed	Thurs	Fri	Sat
		Sun: week of	Mon	Tues	Wed	Thurs	Fri	Sat

Parent(s)/Caregiver(s) who completed calendar: ______________________
IMPORTANT!! SAVE!! KEEP THIS PAGE AND GIVE IT TO YOUR SERVICE COORDINATOR!!

EIP-13 (Rev. 5/06)

Sample page of a Parent Progress Report used by Early Intervention therapists. Published by the New York State Early Intervention Program, 2006.

used by some of the parents in my study. According to the Strivright website, they send EI therapists to the home to provide auditory training approaches and therapies; they "continue to make miracles happen every day" (Strivright Auditory Oral School of NY 2014). According to the National EI Longitudinal Study (NEILS), in-home speech therapy is New York State EI programs' most common form of offered therapy (NEILS 2007, 2). This means that Individualized Family Service Plans (IFSPs) primarily focus on speech therapy and on having parents work with their child to develop auditory skills. IFSP activities may include things like eye tracking behaviors, sound localization, and speech production patterns, which the following two chapters address in more detail.

Specifically, Sonya told me, the training plans and progress reports were developed by EI programs as a way to "monitor what was happening in the home." The reports provide the state with information about the child's progress and parents' compliance with the plan. This monitoring yields information that is also directly available to clinic staff, so they are able to stay current with what happens at home after the child leaves the clinic. Because the plans are so parent centered, Sonya said, "very often having kids in EI helps people to get the implant because they're in the therapeutic mode. They're connected to us."

EI therapists who communicate with Sonya help socialize parents into their new social role as auditory trainers and enforce this therapeutic mode. Sonya, in her dual role as employee of the center and EI coordinator, explained the significance of this web of cooperation in a simple statement: "They trust us." Whether they are at the beginning of diagnosis or months into their EI services, nothing is more important to parents than staying connected to EI and to the clinic. She continued, "They're in the loop. They're in the waiting room, they're thinking about it. They come here for an EI meeting, and they're in the waiting room, and they see a kid who's implanted. Or they come to a parent [support] group, and they weren't going to implant the kid, and they see a parent who brings a child and—'Oh! Look how good the kid is doing, can I have your phone number?' You know how things work in the world when people are looking for information . . . and that's exactly what happens in this culture. This takes time; this is not quick. This is not quick . . . But I think culture is what happens. I think there is [CI culture]. I've never seen articles with the word. But I think it exists, it just hasn't been labeled yet."

In this stage of initiating intervention, parents are socialized to take on an active role, with attendant numerous expectations. They are expected to actively participate in EI-recommended therapies consisting of specific, condoned actions promoting speech and auditory training, and to demonstrate that they do so on a daily basis. But threaded through all of this is also a particular ideology—indeed, Sonya raises the possibility that it is culture—that infuses these efforts to ensure parental compliance. And compliance is central here; Sonya noted that she finds out if parents are not complying and tries to ascertain whether they are just being "resistant." This kind of shared, collective making of what parenting a deaf child looks like (and what being a good mother looks like) is actively and purposely cultivated between professionals and parents.

At the same time, these anticipatory structures located in the clinic soothe parents, minimizing their anxiety. In EI programs, providing emotional support and socializing parents into intervention techniques are intimately and explicitly connected. One reason is that being put into a collapsed patient/parent role when you have a child with a disability is a specific experience. Parents of children with disabilities have been documented with higher anxiety levels than parents of children without disabilities, and thus access to emotional support is crucial (Ingber and Dromi 2010; Leiter 2004). Offering informal support groups and encouraging parents to take active roles in their child's treatment—specifically through participating in EI techniques taught to them by an EI therapist—have been shown to mitigate parental anxiety and stress (Moeller 2000; Sass-Lehrer and Bodner-Johnson 2003).

Every parent I interviewed talked about the positive and pivotal role the EI therapist had in the home, as well as how they came to depend on the therapist as a source of information and comfort. Becky raved about her therapist, who worked with Amy at home "for years. . . . She was phenomenal, and she also had worked with a lot of implanted children, and she had recommended [Amy's school]." Ultimately, the EI therapist's recommendation ended up leading to Amy's educational placement once she reached school age. Becky said, "I gave them a call, I went down there, and they were phenomenal. They were a wealth of information." This is not unique; all of the other parents relayed similar stories.

"It Wasn't Therapy for Him. It Was Therapy for Us."

According to Carol, Sonya helped make the process of instituting EI smooth, although Carol found dealing with the agencies that contract with the New York State EI program, like Strivright, a little more difficult. "With Sonya, you've got to get all this therapy, and everybody has to be approached . . . social services, EI, etc. Sonya took care of all of that; she would set up every meeting, and she's super organized." Sonya helped Carol get EI services through Strivright to work with Jeremy. Carol recalled, "We looked to Strivright; Sonya said you can get speech therapy in the home or where he goes to day care, but we looked to get it in the home." Carol and her husband were, however, interested in having his services at a day care;

they wanted Jeremy to be able to receive services while socializing with other children. She said, "It was hard, because deafness has a lot of other problems. For example, if somebody's bothering him, he can't tell them to go away. So you get some other behavioral problems." Plus, she explained, even though there is a variety of agencies that contract with EI, these agencies may not travel to all areas of the city. She remembered one such interaction with Strivright: "In order to get you as the client, they say, 'We'll go here, we'll go there, we've got people. Yes, do us.'" But once she signed on, they told her, "You're signed up as a Strivright client. Now, let's get you the therapist. No, I'm not going to [day care]. I can't do Tuesdays. Blah, blah, blah."

But, she said, it is complicated: "These are all independent contractors, and I can't blame them." Even though EI had approved therapy in day care, it never happened. Once the therapist tried to come to Carol's home; Carol became very unhappy with the scheduling difficulties. Ultimately, in contrast to the other parents in my study, she chose to go to the center and work with Gretchen, the speech pathologist at NYG, who also ran the support group. Although EI ended up not taking place at Carol's home, she still felt satisfied with the experience because she maintained strong ties to the center through frequent visits:

> Honestly, when he was a toddler, it wasn't therapy for him. It was therapy for us. We watched them play with Jeremy so we could learn to play. And, in fact, what I wound up doing is, my mother would come. My mother would come with me on Fridays to go see Gretchen. And that was one of the best things I ever did, because my in-laws, as much as they love Jeremy, he's just their world . . . they were devastated by the fact that he was deaf, which, actually, really angered me. But, you know, him being defective really bothered them. They couldn't get over it. . . . They were convinced that he wasn't going to be able to get a job when he grew up. . . . Making them go to [Jeremy's] therapy too was the best thing I ever did, because they drank the Kool-Aid.

Carol and her family worry about the future but experience relief through EI services. Sonya alluded to this process as well; she talked about how through experiencing speech therapy with EI service providers, parents become more accustomed to the therapeutic mode and "[settle] down" emotionally. But Carol, who later talked about her use of sign language at home and conflicting feelings about the strictness of auditory training, also seemed to see it with a degree of criticality, understanding it as "Kool-Aid."

According to Sonya, there is a high correlation between parental level of involvement with programs like EI and the likelihood of parents staying engaged for the long-term in the implantation process. One reason for this correlation—at least at NYG—is due to Sonya's dual responsibility: enrolling and socializing parents into EI because of her role as Service Coordinator for New York State, and providing emotional support for parents through their anxiety and grief because of her role as NYG's social worker. This creates a high level of cooperation and coordination between the clinic and the state. Since the emotional experience is so hard, Sonya emphasized to me, counseling must go hand in hand with enrolling families in EI. She went to great lengths to ensure enrollment. She said, "There's some people I've been phoning for five years and they keep calling me back . . . so you have to be, as Sharon says, tenacious. You have got to hold on. . . . My theory is, start with EI, have this nice therapist come to your house, it's very nonthreatening. She's going to play with the baby, but she's really doing [spoken] language stimulation. So we get them in EI."

Minimizing parental stress and anxiety is not the only reason that referral to and enrollment in EI programs is of utmost importance from a clinical perspective. Because ongoing therapeutic efforts have been shown to increase the efficacy of CIs (Geers, Brenner, and Tobey 2011), EI serves as an entry point into training parents to work with their children in particular, clinically significant ways; that is, EI socializes parents into the kinds of interventions they should do with their child. In fact, "in the case of children with hearing loss . . . parents with high motivation have been found to participate in their children's early intervention programs and to cooperate with professionals" (Ingber, Al-Yagon, and Dromi 2010, 363–64). Thus, while the anticipation of parents' emotional needs is an important part of the center's comprehensive approach, their goal is also to encourage and maintain compliance in the medicalized script of deafness.

Ideology of the Therapeutic Mode

The intervention stage is highly structured and monitored. The most crucial component is getting parents into what Sonya calls the "therapeutic mode." The clinic achieves this by fostering a relationship with parents through EI that is founded on emotional support and medical information, by offering grief support, and by socializing families into a "parent-centered approach," wherein parents assume clinically sanctioned therapeutic duties as a part of the child's service plan. Then clinics provide case information to EI, which in turn also provides clinic staff with case information through therapist and parent training reports. In a previous study of EI, Leiter (2004) concludes that because intervention occurs again and again over time, the "habilitative nature of the EI program emerges as staff implement the child's service plan during regular home visits" (839). This pattern exists in the data I collected as well; as Sonya said, there is no "quick fix." Leiter argues that accumulation of interactions in participating in EI over time constitutes the program's habilitative nature. While rehabilitation is the process of restoring what was lost, habilitation creates an ability that never was—or in the case of dealing with children, one that has yet to be developed.

The habilitation-related duties that parents take on with the deaf children I studied plainly demonstrate scientific motherhood and also explicitly invoke the future of the child; that is, habilitation is bounded to the notion that if parents exclusively use these therapies now, then their child will not be hindered in the future by their deafness. This is reflected in Strivright's philosophy, which is "aimed at enabling our students to go on to live absolutely unlimited, successful lives with all social opportunities, academic choices, and career options open before them" (www.oraldeafed.org). Strivright's approach "focuses on teaching children to speak and hear, not relying upon compensatory skills such as sign language or lip reading" (www.oraldeafed.org). They claim the child will be able to speak, hear, and use spoken language in the future, and that visual methods of communication, like sign language, are "compensatory," and therefore not in and of themselves whole. This orientation to deafness reflects a broader pattern where "scientific knowledge is the foundation for public policy in the US regarding child intervention programs for children with disabilities and informs the practice of professionals who work in those programs" (Leiter 2004, 839). For example, according to the American Speech-Language-Hearing Association (ASHA) and the Joint Committee on Infant Hearing (JCIH), EI services for deaf infants should be provided by professionals with expertise in hearing loss, including educators of the deaf, speech-language pathologists, and audiologists. Furthermore, these organizations maintain that deafness that is adapted to with visual language is not only compensatory but also *unscientific* and therefore *unhealthy*.

Socializing parents into this general ideological basis of EI during the intervention stage of implantation and giving parents scientific information and purpose have been shown to increase their compliance with EI and to decrease "levels of pessimism about their children's future" (Ingber, Al-Yagon, and Dromi 2010, 361). And this is exactly where temporal aspects of intervention most powerfully emerge. Everything is oriented to the future: Divisions of labor and the work flow in the clinic anticipate and respond to the latent emotions of parents, while stressing the urgency to undertake these intervention tasks now for the good of the child's future. They also build on the known link between emotional and informational support, and efficacy and involvement in EI (Desjardin 2005). Rose's (2006) concept of ethopolitics describes how medical interventions appeal to our sentiments and beliefs in order to shape our conduct toward ourselves. I would add here that in the context of scientific motherhood, this concept is extended toward those for whom we give care. Rose argued that ethopolitics encompasses a "moral economy of hope in which ignorance, resignation, and hopelessness in the face of the future is deprecated" (2006, 27). I observed the ongoing commitment required by EI to be actively solicited by direct appeals to parents' beliefs about what is medically sanctioned *and* what would be best for their children's futures.

The intervention stage illustrates a new social realm, both in the clinic and in parent communities, that is redefined by anticipation. This is not just about the anticipatory structures that organize clinical practices and parents' experiences but also about how the task of raising and developing a deaf child is reorganized. Before changes in CI technology, implementation of NBHS, and the lowering of the age requirement for implantation, deafness was not a condition that one could *see* in a newborn or infant. In the past, diagnosis was retroactive, a lens through which parents could make sense of the past, make sense of what had *already happened*. But I argue that the changes to technology and policy that have restructured work flow into anticipatory structures have also transformed intervention into an ongoing, future-oriented, scientific—that is, with medically grounded therapies and data-generating parent progress reports—enterprise.

Beyond Decision Making

The significance of the therapeutic culture created by EI also suggests that instead of there being a onetime decision to implant, this decision is just one step in a larger story. In other words, the decision to implant becomes a logical expression of this therapeutic culture, transforming it from a pivotal, onetime moment into one more natural step in a larger process. Yet many previous studies of implantation focus on the decision making around surgery (e.g., Nikolopoulos et al. 2001; Christiansen and Leigh 2002; Li, Bain, and Steinberg 2004; Okubo, Takahashi, and Kai 2008). This chapter and the preceding one, however, show that families are being groomed for implantation in the months or years leading up to surgery. When I first began my research, I mistakenly thought I would witness a moment where parents decided to implant, or a moment where the CI seemed to work and the transition to being a CI user who spoke and listened well just "happened." But this expectation was not based on what implantation turned out to be: a very long process in which events like the surgery are but one step. In his study, Blume finds what he calls a "blandness" in decision making about CIs, writing that the choice to implant is "more or less automatic for most parents of deaf children in rich western countries" (2010, 171). These are the reasons why I am not focusing on decision making in this book. Clinical studies of implantation to date have mainly focused on parental decision making about surgery, positioning implantation as a stand-alone surgical event rather than a piece of a larger socialization process. These studies mark surgery as some kind of endpoint, rather than one stage in an overall implantation socialization process that is highly monitored and structured.

Parent Support Resources

Another key aspect of this therapeutic culture is the emotional support and networking that occurs between parents. According to Sonya, support groups are one of the most effective tools for both tending to parents' grief and ensuring their compliance regarding enrollment and socialization into EI programs. The CI support group at NYG provides emotional care to parents, fosters parent connections among them, and creates a sense of community. All of the parents I interviewed—and many more whom I observed at additional support groups outside of the center—talked about the helpfulness of not just support groups and organizations with parent resources (namely, the Alexander Graham Bell Association),[2] but also other hearing loss organizations and local school programs. The CI support group at the center is part of what Sonya referred to as the center's "comprehensive approach," which acknowledges that implantation takes a long time and demands attention to the family's emotional life. She noted, "It's comprehensive. What's happening with the whole person, what's happening emotionally? We've had people who were depressed. This [the CI] is not a quick fix." These support groups do not just do emotional work; they also have a cultural role. She continued, "I've seen parents come full circle. In group therapy they say that when you're really comfortable and you're training the next newbie a year later, that you've made the full circle, you've come to terms with it because you're helping the next person . . . you've become enculturated."

The center runs multiple groups, but the parent group stands in stark contrast to the adults with CIs group. "The parent groups become more therapeutic because it's small and they open up," Sonya explained. By contrast, the adult CI group at the clinic tends to be more informational, with speakers from CI companies and other adult CI users coming in to give talks. "They're still a little cautious. They'll say a little bit here and there, but they don't want to come for therapy," Sonya commented. In contrast, she said, "The parent group is really more therapeutic. People will cry, they'll be sad."

Gretchen, one of the speech pathologists at the clinic, has been working there for more than fifteen years. She runs the CI support group for parents and indeed described it as "heavier" than the others. In the parent group, she observed, "people will express their personal feelings and their grief, and all their emotions, and that'll be much deeper." The parents who attend the group are those whose children have had audiological testing at the center, and they may be at various points in the process. "Some of them are undergoing the core evaluation for Early Intervention. Some of them are having to make decisions about whether they want home-based or center-based therapy," Gretchen said. Gretchen and Sonya take a team approach to the parents: Sonya speaks to many of them either in person or on the phone, then sends Gretchen a list of parents needing additional support—the ones Gretchen should be on the lookout for at meetings—and their contact information. All of the parents, Gretchen said, wanted to come to the center's support group but sometimes experienced obstacles. This situation informed the center's approach of staff keeping each other apprised of the parents' emotional states. "[The process of implantation] is really hard, so some of the parents drop by the wayside, which is why Sonya always gives me their names," Gretchen explained. And the parents who are really struggling—those who are facing grief or in danger of not returning—are the ones Sonya stresses to everyone in the clinic to stay on top of. As we were talking, Gretchen's memory was jogged: "Sonya just gave me a name again this morning, somebody that Sharon just saw and diagnosed . . . and then recommended the parent group immediately." She rifled around on her desk, looking for the paper.

Gretchen has crucial help running the support group. She started this group many years ago with Nancy, a parent whom the clinic has dubbed "the old-timer." As I spent more time at the clinic and in schools and parents' homes, the active and central role that Nancy plays in so many people's lives became clearer and clearer to me. Indeed, she had been the driving force behind the creation of the support group more than a decade earlier. According to Nancy, in the late 1990s, when her daughter Anne was implanted, the center "didn't provide what we needed and that's why we started the group. . . . There was a lack of support. There wasn't enough information going around." So, in an effort to make the CI process easier on parents, Nancy and Gretchen started the group. According to Nancy, this was a strategic move. Since Gretchen is a professional with a certain role in the clinic, Nancy explained to me, she "may not be able to tell parents certain things." But, she said, the great thing about a parent group is that "as a parent, I can tell them anything. We'll be gentle about it. But there comes a point when you say look, getting the implant for your child is probably going to be the best thing you ever did, and you know what, if it doesn't work out, you still tried the best that was available."

So, even though parents in the group are in various stages of the process—for example, coping with diagnosis, or just about to face implantation surgery—they all have in common the fact that their child has already been identified and they are receiving EI services. When I asked Nancy about this process of moving through the stages, she admitted that some parents struggled. However, she said, "It's only because they are ignorant of the facts." The problem, as she sees it, is that "they assume that if you're deaf, you have to sign. We give them the statistics. More people who are hard of hearing or deaf are oral [use spoken language] than are using sign." Parents' response to this fact, Nancy said, is often surprise. She said, "That's why I love bringing people to this!"

Once at the group, Gretchen told me, parents have "the opportunity to tell their story." Many of the parents I interviewed attend or attended this support group at one point, and used other support groups and resources as well. For example, Carol joined Hear Us Long Island, which is closer to where she lives. "I met mothers there. I talked to them. I needed to go to something where I could meet parents," she said. As for the support group at the center, she went there much later: "You should be supportive [of other parents] and show up. So we try and go." The emotional support overlaps with information sharing. One time, Carol recalls, the group started talking about sports. "You find out how to keep that magnet on his head, if he's going to be playing sports." Parents also discuss the complicated educational system they must navigate. While Sonya provides them with a lot of information on that subject, all the parents told me how beneficial it was to talk to other parents. Carol told me, "It really is a kind of networking effect; that's the biggest thing. . . . Me and these two other moms got to be friends. We formed our own little girls' night out, which was, basically, you know, we've got to get together and talk about the kids."

Coping with Anxiety and Guilt

One afternoon, I arrived at Jane's for our first interview while her daughter, Lucy, was at school. She answered the door while balancing pieces of electrical wiring and a light fixture in her other hand. She was a little frantic, cleaning up as we walked down the hallway to the living room. I asked her how she was doing with everything; Lucy was supposed to have her implant surgery soon, and in anticipation of this, she had transferred school programs from a signing classroom to one that uses spoken language. Due to her specific type of hearing loss, which is progressive, she had not qualified for an implant until age four. Jane felt overwhelmed with everything involved in the process.

The first thing she talked about was her support group and how much it helps soothe her emotional pain. "I understand the process of grief," Jane told me. She is an active support group member, having made the full circle of supporting others that Sonya spoke of previously. She told me that "there's a shift, and you won't even notice it, and you'll find yourself being in support of others." She began attending support groups because "I wanted to know it was going to be OK. And without me being there [at support groups now] and telling these parents it's going to be OK [she drifted off] . . . I make them comfortable, I tell them it's going to be OK."

She began to cry. "I haven't had this in awhile," she trailed off. "What haven't you had?" I asked. She explained to me that she has had tremendous anxiety since Lucy's diagnosis. It had resulted in her having panic attacks and experiencing numbness in her fingers. I worried that I should find her a tissue. As I looked around for one, she said, "I knew this [interview] was coming and I worked myself up a little bit. I tried to distract myself with the lighting fixtures. But it's necessary if it helps other parents. . . . It's good, it's cathartic." She was sitting on the couch, her hands folded in her lap. She looked down for a few moments, quiet. I simply sat with her. Then, she looked straight up at me, tears in her eyes, and said, "It's OK, it's OK to say I have anxiety. Maybe if I speak up, maybe I'll give it a voice, because you know parents are going through it, you know they are. Then maybe they can say it's OK to talk about it. Then maybe they can heal a little faster, and they can be more beneficial to their children."

When I visited another mother, Becky, at home one afternoon, we had coffee and sat in her living room. She talked about how her daughter, Amy, responded when she got the implant approximately a year earlier, at the age of two. Amy had been diagnosed at eighteen months of age. When I asked Becky about some of the emotional struggles over the years with implantation, she told me how Amy would cry and cry and refuse to wear the CI after she had the surgery. She talked about how difficult it was to watch this. "I felt a lot of frustration and pain." She wanted to emphasize to me the "real story" of CIs; that is, she wanted to make sure that I knew that there was a great deal of difference between her experience and what she saw on the CI company's promotional materials and those YouTube videos showing kids hearing for the first time. She dubbed this the false "Hallmark moment," and she was angry about it. She wanted to talk about how CI corporations—and many videos available online—showed the child when the implant was turned on: They turned and looked at the speaker, at the sound of the speaker's voice. For Becky, and many other parents I spoke with, this felt like false representation. At worst, it was manipulative and, at best, woefully oversimplified. It was in the support groups, Becky said, that she found out that other parents feel this way too and share her experiences. Other parents do not have that moment either; they said, "Oh, I didn't have that and it's OK."

But Becky also needed to share some of the day-to-day battles she fought with Amy for months. Amy threw tantrums and had trouble adjusting to the feel of electrical current from the device. This caused Becky to feel terribly guilty, and she struggled for years with these feelings of guilt. These were extremely difficult experiences because she questioned herself and wondered: How do you know what you are doing is right? These experiences kept her needing ongoing support, so she attended support groups at the NYG and also at Amy's school. She described them as "absolutely helpful." Parents told her, "It's not just you," assuring her she was not alone; apparently many families deal with this same situation at home, but no one in the clinic really speaks about it. Becky commented, "The main thing that I got from other parents is: 'We understand, we went through this. Try this, try that.' Or they'd say, 'This is what we did when my son or daughter didn't want to really wear it.'" She also spoke of the powerful camaraderie—and the need for it: "Let's help each other out. Let's be there for each other. Emotionally it helped me out a lot." The main reason, she says, is "because I wanted to know what to expect next."[3]

Parent and Professional Networking Events

On a cold January evening I headed to a parent support group hosted at a local school and sponsored by the Alexander Graham Bell Association (AGB). AGB also holds informational seminars and hosts various speakers on CI-related topics, but on this night, the organization was explicitly holding an informal support group for parents to talk about anything they wanted. When I walked in, I found Nancy, the old-timer, greeting people. All kinds of publications were sitting out on a table, such as *Volta Voices,* published by AGB, and other materials from CI-related conferences. There was a food spread and multiple tables. Slowly, parents trickled in and stood around informally in a circle. Some were there alone, others with their spouses. They timidly introduced themselves, asking each other what degree of hearing loss their children had, if they were implanted, and where they went to school.

Once I sat down at one of the tables, I was joined by a couple who have a daughter who is deaf but was not implanted. According to them, she does not speak well. As we were talking, I realized that in addition to parents, a lot of clinical and educational professionals were in attendance, because I recognized them from other meetings or events. As people got something to eat, milled about, and sat down and talked, the volume in the room went up. There was no formal structure to the meeting, but the professionals in attendance split into pairs, joined different tables, and sat down and started conversations. People became more comfortable and opened up more as the evening wore on. A speech therapist who had joined my table spoke about her work. When referring to a young implanted girl she currently worked with, she talked about when the child was deaf, as if her deafness was in the past tense, as though it had gone away because she was implanted and had learned to speak so well. The speech therapist then quickly corrected herself; she said that the child actually was deaf, but that she had forgotten, because the child functioned just as a hearing person would. She noted that some of the parents in the room had children that she had worked with in the past. She pointed to another family in the room and said that she had worked with their child as well. She offered to put the parents at the table in touch with them, assuring them that the community was really small. Other parents nodded in agreement that it was like a little universe.

Farther down the table, I heard one woman talk about how frustrating it is that people have a really good idea of what Deaf culture means because of media coverage. But, she said, people do not know what it means to have a CI. A few times during my fieldwork, I encountered a similar positioning of Deaf culture as the dominant idea of deafness to combat. Many parents and professionals routinely expressed frustration that people often assume that deaf persons should learn sign language. This is one of the ways Deaf culture is a sort of implicated actor (Clarke and Montini 1993); it is positioned as something to be against, something to defy, something from which to differentiate. I thought about how when I attend Deaf events, often the same thing happens that happened that night at the CI support group. People come in and hug one another. Introductions center around where you are from, what school you attended, whom you know in common; the community is rich and expansive. But in the Deaf world, community members lament steady rates of implantation as indicative of a belief that deafness should be corrected and that d/Deaf people should not use sign language. From the Deaf cultural perspective, the medicalized idea of deafness (epitomized by the CI) is the dominant understanding—and this is what they want to combat. Either side perceives the other to be dominant. And furthermore, many (but not all) see these worlds in stark opposition. But do these communities need to be so separate?

As I sat with the parents who attended Nancy's meeting that evening, people asked each other about the degree of their children's hearing loss and whether they had a hearing aid or a CI. The couple sitting next to me talked at length about the emotional aspects related to dealing with their daughter, the one who was not implanted and did not speak well. The father explained that his biggest fear was that his child would be alone. The mother, with tears in her eyes, told me how her daughter kept asking her why she was not "normal." As the meeting ended, she turned to me and said that her daughter did not speak well, did everything through the computer, and had a lot of social relationships though text and e-mail. Both parents seem worried about the authenticity of such interactions. But mostly, they were worried about their daughter being alone—a particular type of "imagination work" that people often engage in regarding people with disabilities in general but especially regarding deafness.

Living within This Therapeutic Culture

In the past, deaf children were often not identified until past infancy. But today, deaf children are identified so much earlier that now an entirely new set of opportunities to intervene is available, and those opportunities have been formalized through a variety of interinstitutional cooperation and anticipatory structures. These opportunities, however, also inundate families with specific duties, responsibilities, and forms of labor that are legitimated through a medical perspective. The parents in my study use these resources and consume the services offered in the clinic and elsewhere. They are expected to integrate these practices into their homes and to use these services to maintain their emotional health and cope with the grief, stress, and anxiety that can accompany a diagnosis of deafness. These highly structured duties form a therapeutic culture that provides parents with emotional relief and support. They are kept extremely busy, and their progress is highly monitored; this medical script of deafness is so deeply institutionalized and consistently reinforced that it also produces what one might call a community that serves parents' emotional needs.

This community—or, per Sonya, "CI culture"—is primarily a therapeutic culture. Coming to terms with having a deaf child was painful for the families in my study, but part of the relief for parents occurred because of the anticipatory structures surrounding deafness and the technology of the CI. The resulting therapeutic culture gives parents hope for the future while simultaneously appealing to their anxieties about it. They are socialized into an ideology of hope, where what Sonya referred to as "staying involved" and being "on top of things" requires ongoing and intensive labor that attempts to render invisible in the future the child's current deafness.

This culture is highly boundaried; there are strict borders between this therapeutic culture, dominated by a medicalized script of deafness, and the Deaf cultural script. These borders appear in discourse about what it *really* means to be deaf and about which version is more dominant: Nancy focuses on exposing parents to the reality that they do not have to use sign language, while other parents and professionals worry that the Deaf cultural script receives more media attention. But regardless of which script of deafness is "right," each side perceives the other as being in opposition and as being the dominant script to push back against. This kind of border work solidifies the therapeutic culture surrounding the CI. The culture has been built not only through the institutionalization of the medicalized script of deafness but also through shared experience, membership in the resulting therapeutic culture, and a sense that one is struggling to be seen as a legitimate version of deafness.

Finally, the patterns I observed further illustrate ambivalent medicalization. Parents socialized into a therapeutic culture are both indoctrinated into a particular ideology about deafness *and* experience relief, support, and purpose. Something is gained, while other possibilities of seeing deafness are lost. The therapeutic culture expects mothers who are raising deaf children to employ a particular set of techniques; mothers who do not participate in the scientific motherhood endorsed by the clinic are labeled "resistant." Leiter (2004) argues that there is something "problematic" in putting these parent-centered programs into practice. She refers to it not as a therapeutic culture but as a *therapeutic imperative*. But this imperative also coexists with something else, with parents' drive to feel relief and to belong to something larger than one's own experience of having a deaf child. Paying attention to these "soft" aspects of intervention, such as the anxiety of parents and the nuanced role of emotion, reveals both an imperative and a reward, or a Foucauldian practice of care for oneself that is extended to one's child (Foucault 1988).

Ambivalent medicalization attends to the capabilities of medicine and the emotional relief. It also points out the labor imperative and the resulting "subtle restructuring of patients' or professionals' identities" that come with new technologies and practices (Timmermans and Berg 2003, 104). The availability of the CI not only reconstructs deafness as a neurological problem—to be examined in chapter 4—but also changes families' experiences due to shifts in state policy that make the device more available and initiate earlier and more organized interventions to anticipate the device. As the next chapter explains, class position affects how professionals work with families and how families move through intervention and into the next stage of implantation.

3

CANDIDATES FOR IMPLANTATION
Class, Cultural Background, and Compliance

> *For middle-class mothers, the boundaries between home and institutions are fluid; mothers cross back and forth, mediating their children's lives.*
>
> • Annette Lareau, *Unequal Childhoods: Class, Race, and Family Life*

Throughout the first two stages of implantation—identification and intervention—parents are socialized into the therapeutic culture of the clinic. The next stage of implantation is transition into candidacy. In this chapter I describe the determination of candidacy, the bureaucratic steps involved, and the preparation for CI surgery. I include interviews with parents showing how they understand and experience the tasks related to candidacy, as well as data showing how audiologists and other clinical professionals assess parents' compliance in these tasks.

The most significant features during the transition to candidacy stage are the powerful role of *social criteria* in determining candidacy, and the part that class and/or cultural background plays in whether parents satisfy these criteria. Although the child's audiological status and ability to meet insurance requirements are a crucial part of determining candidacy, professionals also evaluate children socially. This social evaluation is largely based on parents' behavior and resources, which are influenced by class and cultural background. Because the therapeutic mode established in intervention carries over throughout the rest of the implantation stages, parents are expected to have been engaged and continue to engage in sustained therapeutic labor. Their consistent, ongoing compliance is expected in two main ways: following audiological recommendations of hearing aid use, and conducting the auditory training suggested in the EI service plan. Professionals also consider the parents' emotional state, which they largely frame as parents "accepting reality."

The extensive, invisible work done by parents as they reach for candidacy and prepare for surgery requires resources. Currently, there is a dearth of information regarding either the socioeconomic status (SES) or race of children who receive CIs, as well as a lack of understanding of the distinct effects of class and race as individual variables. For example, twelve-month-olds are documented as the most rapidly growing population of those receiving CIs, but there are no comprehensive data on their breakdown by social categories (Belzner and Seal 2009). However, Belzner and Seal, citing one study of implanted children under the age of eighteen (Stern et al. 2005), summarized the existence of racial stratification: "A higher proportion of White and Asian/Pacific Islander children and a disproportionately low number of Black and Hispanic children receiv[ed] implants" (2009, 313). Stern and colleagues (2005) reported that white children are implanted at a rate three times higher than Hispanic children and ten times higher than black children. Additionally, it is noteworthy that most, though not all, of the audiologists in my study are white; data show that more than 90 percent of EI providers are also white, female, and monolingual (in English) and have an advanced degree (National Early Intervention Longitudinal Study 2007)

All of the families in my study are also white and middle class, a sample that seems to reflect an overall trend, but without more data, this trend is unverifiable.[1] When I attempted to find demographic data during fieldwork to analyze the populations of children implanted at NYG and those implanted nationally, Sharon indicated that such data did not exist. In subsequent inquiries to Cochlear Americas, I was also told that these data were not available. According to Belzner and Seal, "Very limited attention has been given to race and ethnicity, socioeconomic status. . . . [instead there's been] far more attention on the communication modality over the years, namely whether children use sign or speech" (2009, 313). (This focus on language outcome runs through all stages of implantation, and I will more closely examine it in the next chapter.) Belzner and Seal conclude that although very little is known about the "influential dynamic of socioeconomic status . . . [studies] are suggesting that

SES is important in outcomes' variability and measures" (2009, 311). For example, studies of children with CIs in the United Kingdom "showed that the higher the SES, the more likely the implanted child used spoken language" (Belzner and Seal 2009, 330).[2]

I observed that the social criteria used to determine candidacy for implantation were influenced by the family's class position and cultural background. However, because I worked with a small sample, this link is difficult to generalize. Nevertheless, it is useful here to reference previous sociological studies of social class and families' integration with formal institutions. In the past, a strong affinity has been found between the goals of formal institutions in society (of which the processes of medicalization and the therapeutic mode I describe here are examples) and the ethos that characterizes a white, middle-class parenting style (e.g., Blum 2007; Lareau 2003; Francis 2012). This alignment results in certain families being more willing to adhere to sustained therapeutic labor and having the resources to do so. It also affects how audiologists and other allied professionals evaluate parents and how compliant they see them to be. As a result, class and/or cultural background may have a bidirectional impact on who becomes seen as a candidate for implantation.

In one direction, middle-class parenting style aligns with formalized institutional social recommendations. Lareau (2003) outlines this phenomenon in her study of the links between inequality and parenting styles. She found that white, middle-class parents see themselves as actively developing their children and engaged in what she calls "concerted cultivation," a style of parenting where parents "actively fostered and assessed their children's talents, opinions, and skills" (Lareau 2003, 238), especially through organized activities. And in the case of implantation, I would argue that families engage in concerted cultivation through *organized therapies*. There is a strong affinity between the logic of concerted cultivation and medicalization, something Blum (2007) identified as concerted medicalization.

Because institutions in society "firmly and decisively promote strategies of concerted cultivation in child rearing . . . families that do not adopt a concerted cultivation approach tend to feel a sense of distance from such institutional experiences" (Lareau 2003, 3–4). All of the parents in my study had been recommended to me as participants precisely because the audiologists deemed them compliant:

They were perceived to have consistently integrated the clinic's and EI therapists' recommendations as they readied their child for implantation. In short, they willingly practice particular methods of intervention from outside institutions, namely, the clinic and social service agencies, in the home. They took to the therapeutic mode and integrated it into their parenting routines.

In the other direction, however, audiologists' perceptions of parents' behaviors lead them to make assumptions about the viability of the child's candidacy; that is, judgments about their class or cultural background may perpetuate inequalities in implantation. These assumptions are not entirely unfounded, as it is well established that higher participation in EI services is associated with higher rates of implantation and better outcomes (Niparko et al. 2010). But in an earlier survey study of audiologists in the United States specializing in pediatric implantation, Kirkham and colleagues found that audiologists identified an SES-related disparity and overwhelmingly "perceived an effect of SES on post-implant speech and language outcomes" (2009, 516). In qualitative responses, audiologists "uniformly demonstrated . . . that lower SES patient populations were more likely to experience reduced speech and language outcomes" (Kirkham et al. 2009). The study's authors give two primary reasons to support their perception: (1) parental "shortcomings," such as parental self-efficacy and adherence to recommended interventions, and (2) other external factors such as a lack of resources or access to therapies (Kirkham et al. 2009). These kinds of assumptions can result in a child *not* being deemed a CI candidate. Class and cultural background may manifest in the extent to which parents can or will engage in the invisible and ongoing labor required of implantation. In this chapter, I show exactly what this assessment of parents' shortcomings looks like and how this shared ethos of concerted cultivations comes to dominate assessments of whether a child is a viable CI candidate.

Finally, parents' immigration status or cultural background potentially conflicted with audiologists' imaginations of ideal CI candidates. If another language besides English was spoken at home or parents adhered to cultural norms from their home countries, this was seen as a deterrent to adequately socializing them into the script of medicalization. Thus, families who "had culture" were potentially seen as problem parents. Yet audiologists were not aware that the medicalized script of deafness and the professional practices in audiology and other associated fields are also cultural and organized around collectively agreed-upon norms. The culture of implantation, of the clinic, and surrounding the technology—which Sonya so aptly referred to as CI culture in the previous chapter—is cordoned off as scientific and objective and thus in opposition to culture. This opposition hearkens back to the competing narratives on deafness and demonstrates that "culture" as a larger category is operating here; that is, the assumption that Deaf culture is incompatible with the CI is also applicable to other minority cultures.

Integrating Sustained Therapeutic Labor

In working toward candidacy, families are told to focus on two important steps: implementing EI and/or other medically sanctioned educational recommendations over a sustained period of time, and putting hearing aids on the child. Once families enroll in EI, they are able to access an assortment of services and therapies as they acclimate to the therapeutic mode, and report these to their EI service provider as part of the "parent-centered approach." They may have not only established a relationship with an EI therapist but also enrolled in infant/toddler programs at local schools with specialized education programs. In the case of Morgan's family, their relationship with their EI therapist and participation in school programs were still continuing some four years after diagnosis. Long-standing parent-professional relationships like this are common in this sample.

I first met Morgan early on in my fieldwork at the clinic. He had been implanted at age two and a half. When I initially met him at age four, I was struck by how well he passed as a hearing child; had I not known he had a CI, I would have assumed he had been born hearing. In late August, I visited his home and spoke to his parents about his EI services. This was the first and only interview where both parents were present. As we sat in their living room, Julia told me how wonderful their EI therapist was: "She was really good as far as training . . . telling us what we needed to do." Paul added, "Yes, her name was Marianne. I still have her number in my cell phone because there's still so many people in my parish that say, 'My kid is delayed, do you know anybody?' I'm like yes! I'll scroll down and, you know, I'll always use her. I'll always recommend her no matter what. Marianne was fantastic, still is fantastic."

Julia nodded in agreement while Paul continued. "Marianne came when he was three months old and started working with him. We also went to the deaf school at three months old." They both went on to tell me how much work was involved, with the high number of phone calls and the time spent doing research in order to comply with all of the demands of implantation. But their EI therapist, Marianne, had been a key source of information for them. "She told us about [deaf school program]. She actually gave us names of other parents who were fine with her giving their names out. We spoke to them," Julia added. And upon Marianne's recommendations, Julia and Paul enrolled Morgan in the school's infant/toddler program. "It all went very quickly."

Julia had also tried to go to support groups, but she found all the parents to be speaking a lot of medical lingo she did not yet understand: "What type of hearing loss it is, and what type of genetics it was. When they were saying Connexin 26 [the gene associated with hereditary hearing loss] and we haven't even done our genetic testing yet. We didn't know what they were talking about. We just felt like between the speech therapist and [the school], they were educating us so let's just stick with this for now."

Both parents also attended workshops at the center. NYG periodically brings in a speaker, or a "Cochlear representative," which is usually a "successful" adult CI user who may be employed by the company, Cochlear, which manufactures CIs. Paul explained, "Before we had the surgery, those different workshops that we went to really helped us to understand." Julia and Paul met one of these representatives at a workshop at the center prior to the surgery. As Paul told it, "Both of us met a guy who was thirty-six years old, and he was a Cochlear representative and he himself had CI." The workshop was also the same day that Julia and Paul met Dr. Brown, Morgan's surgeon, for the first time. Paul reflected on this: "He got up there, and he spoke and everything. He was 100 percent for cochlear. Wow. I spoke to him for a few minutes, and then I basically took all my time and gave it to Dr. Brown because I just thought Dr. Brown was very easy to talk to. I was blown away. He's doing just fine, and he's talking and he's doing great! [The representative's] speech was fine. That made me feel really good to see that." Meanwhile, the center had helped Julia get all of the EI services in place, and Marianne, the in-home speech therapist, was coming to the house twice a week "to be with Morgan" and to go over all the information. "Then fast forward right to his CIs, and he has continued on from three months old right through now at [the deaf school]," Julia said.

Paul leaned forward to tell me his past routine of accompanying Morgan to the deaf school program on Monday mornings. "I remember going to [the school] at nine in the morning and being there until eleven. Because, see, it's not just them doing it; it's also teaching the parent that this is what you do." He began to comment on what these therapy sessions were like, turning to the issue of gender. He emphasized that he was always the only father in the group: "I'm not like a mushy gushy kind of person. When I'm surrounded by women . . . women acting all mushy, gushy, I have a hard time with that. I openly admit that, I do. I have a hard time . . . I didn't want to be mushy gushy. I couldn't do it. So after awhile they said, 'Oh you have to do this, Dad.'" Julia started laughing; they talked about how Julia would ask him how the sessions went when he came home afterward. Paul remembered, "I said it was so horrible. I had to do this stupid song about the teddy bear."

Paul recounted all of the women's names, marveling at how supportive of parents the in-home as well as center-based EI therapists were. He especially remembered one woman, Kate, as particularly enthusiastic. "She'd run down to the cafeteria and she'd run back. 'Here's coffee with milk. Please drink this. You've got to stay with me. I know you're exhausted.'" In that program, he spent time with both a speech therapist and a "teacher of the deaf." Julia piped in, saying to Paul that all of the professionals were "good at pushing us, like they did with you and the coffee, to push you beyond your comfort level. [Julia turned to me.] It's not about your comfort, it's about what [Morgan] needs."

Julia and Paul willingly accept their prescribed roles, and they have the time and resources to engage in all the therapies, as well as the know-how required to create relationships and network with those in a position to help them. They network with others, consume and integrate a variety of interventions and services, and, most importantly, see the task of "working with" Morgan as not just the job of professionals but also their job. Lareau (2003) emphasizes that middle-class parenting style correlates with attitudes toward institutions; in white and/or middle-class homes, there is more of a "seamless overlap" between institutions and the home. Chapter 4 documents some of the moments challenging this seamless overlap.

Paul was unusual in the clinic, in that the task of working with Morgan in a therapeutic way transcended his understanding of his gender role. Before I met Julia and Paul for the first time, audiologists in the clinic took me aside and told me that I would be impressed with him. Paul was "one of the good ones," they said, for taking so much initiative in Morgan's care. They loved it when he accompanied Morgan to the clinic, and they directly praised him for it. I never saw such praise for mothers who brought their children consistently, while he was deemed an outstanding father for doing so. In general, this gendered pattern of care is so prominent in the clinic that it feels unremarkable when mothers are there but utterly remarkable when fathers are. The only time fathers tend to show up in the clinic is just before and during surgery. The clinic staff seem socialized to this norm; I never heard them critiquing fathers' lack of participation. These gender norms come up again in Jane's story below.

Hearing Aids and "Candidacy Anxiety"

During the course of these ongoing interventions, one of the main factors used to determine a child's CI candidacy is his or her measured benefit from hearing aids. While parameters for CI candidacy continue to evolve, most insurance companies require not just a certain degree and type of hearing loss but also documentation that hearing aids provide insufficient benefit. Dr. Brown, the surgeon at NYG, emphasized that you must first see "if they're getting inadequate benefit from a hearing aid, and how you define inadequate benefit really depends on the age of the patient." As such, audiologists recommend hearing aids to facilitate auditory training and to provide amplification. Specifically, the child must be provided hearing aids, and his or her response to them must be monitored during EI's home auditory training activities. Dr. Brown explained that "a hearing aid is first-line treatment," but CIs are not. With all the talk of implantation and with introduction to the idea of CIs happening on the same day as diagnosis, most of the parents in my study regard the hearing aids as a "hoop" to jump through for insurance companies in order to qualify for the CI.

As parents move along in the intervention phase toward candidacy for the implant, what gets evaluated is the adequacy of hearing aids, along with parents' performance as auditory trainers and overall participants in the therapeutic mode. Given the age range of potential CI candidates—from newborn to later infancy to even five years of age—there are key milestones to look for in order to determine how to proceed. Dr. Brown commented, "If it's an infant, say they aren't making any response to sound . . . or they're not beginning to verbalize. You look at whether they're babbling. If they're a pretty young infant or an older child, are they developing any language skills at all?" Regardless of the fact that each stage of infancy involves a different relationship to sound and language development, all parents and audiologists use hearing aids' lack of efficacy to determine whether a child is a CI candidate. In other words, candidacy is predicated on a *lack* of access to sound and language, despite appropriately using hearing aids.

The parents I interviewed reported that they were anxious to proceed to candidacy; they wanted the time span between intervention and surgery to be as short as possible. Because of the need to move through the stages of implantation, however, parents expressed a constant worry about the "wasted time" between diagnosis and implantation. Consider Jane, who had worried about Lucy being a "late implant." Lucy, unlike the other children of the parents in my study, who averaged under two years of age, was not implanted until she was four years old. The delay was caused by difficulty in diagnosing Lucy's type of hearing loss. This had prevented approval from the insurance company and caused some hesitance on the part of the surgeon and audiologists. "What worried me the most is Lucy is a late implant, and her speech is going to be . . . I just keep seeing all these one-year-olds [with CIs] [trails off] . . . The last time we saw Dr. Brown before surgery, I said, look, every time I come into your office my heart breaks. I see these little kids that are barely walking, and they're getting CIs. I choked up and said, please, please [putting her hands together] let us be next, she's four! And every day that goes by is another day that we sacrificed her language. It's more time that we've lost. It's already late. Shit or get off the pot already."

Similarly, Carol also told me that once she and her husband learned about the CI, at the time of Jeremy's diagnosis, she had decided that the CI needed to happen instantaneously. He had turned to her and said, "Well, we've got to get this cochlear implant." Becky, too, recalled being told about the CI right after her daughter was diagnosed, and feeling sure about wanting it as soon as possible. Nancy, the old-timer, had also been ready for the implant from the moment she and her husband found out about it. But Jane had a long wait between diagnosis and candidacy; she even had to endure Lucy's being deemed a candidate and then having her candidacy revoked. Lucy's particular kind of hearing loss is caused by Pendred syndrome, a genetic and progressive form of hearing loss that usually appears just after birth. The hearing loss often occurs suddenly and in stages. Jane remembered, "Pendred is a rollercoaster. You're up, you're down. And I remember Annette, the chief audiologist at NYG, saying, 'I've heard of this, but I've never seen it.'" There were times when Lucy seemed to have her hearing: "I could call her name, and she would respond with no hearing aids." And then suddenly, Jane said, it would be gone. "And I knew when it was gone. At this point, all the doctors at the center knew. . . . They didn't want to touch her. 'Let's watch her for a while.' Watch her for a while?" She threw her hands up in the air in exasperation as she told me, "How can Lucy learn language if one day she wakes up and I don't know if she's hearing that day?"

Even more frustrating to Jane was that a second auditory brainstem response test (ABR) confirmed the extent of Lucy's hearing loss. Jane was ready. "She had auditory memory;[3] I knew it could be salvaged." And she felt that her case would be heard. She felt that she had a strong relationship with the center. "We were there every week because she ate the [hearing aid] mold; she found ways to destroy it." And so Jane did all the things she had been told to do for Lucy; she got EI services at home, and she made sure Lucy wore the hearing aids. She took out a color-coded calendar notebook with various colors shading the days, each one representing a different type of therapy. She felt extremely effective with getting "tons" of EI services, even if she had to be confrontational to do it. "They [EI providers] know me. I'm the person if you need something, it's done. I've always been a good worker. . . . I had to fight for her. I went in, I went in very defensive, I went in ready to start a fight." Jane has no problems intervening on behalf of her child.

But then Lucy seemed to be responding to the hearing aids again, so surgery still slipped through her fingers. It had appeared that Lucy was a candidate, but the testing results were conflicting. Jane got everyone together: Annette, the chief audiologist; Monica, the primary pediatric CI audiologist; and Dr. Brown, the surgeon. "We were ready to do surgery, and then they decided not to," Jane told me. One appointment in particular was extraordinarily stressful for Jane to recount: "We were picking out the color of the processor, and then they were like, let's not do that, and they shut the box. . . . Monica was showing us how it all works, and I remember: She shut the box. [She mimes the closing of a box.] The box was wide-open, we were going over everything, and she said, it's based on whatever hearing test we get today, that will be the deciding factor. Dr. Brown had said that if it's the same or better, we're not going to touch her. But if it's worse, we'll do surgery. And they brought Lucy back into the room, and Annette said, 'It's the same,' and Monica shut the box. We were so close."

At this point, Jane began to cry. "I was mad," she said. She felt desperate to get Lucy's hearing loss treated, and increasingly anxious. She emphasized to me what a "big deal" that appointment had been for her and her family by telling me that her husband "had even taken the day off" to accompany her to that appointment, which is not something that he had done before. "We left the appointment, and he said, 'That was bullshit. I had to show up for that? I was supposed to be signing a release form for her to have surgery. *That's* why I took the day off.'"

Jane continued to fight for Lucy, she told me. "I had a couple of phone calls with Monica. . . . I finally got my wits about me, and I said, you need to figure out what is going on; we need to make a decision. Shit or get off the pot. It's been a year." This time, she also threatened to leave NYG: "If I come in one more time, I'm taking my paperwork and going home." At this point, she said, she became aggressive. The next time Lucy was tested, surgery was a go. And just three months after this interview, I sat with Jane and her husband all day at the hospital while Lucy received a second implant.[4]

Jane was able to get all the services, navigate the health care system, fight for interventions from the state, and feel capable and purposeful at using institutions to benefit her child. This reflects Lareau's findings wherein we believe that "outcomes are connected to individual effort and talent, such as being a 'type A' personality, being a hard worker, or showing leadership" (2003, 7). I see this characteristic in many of the parents in my study, especially Nancy, the old-timer, who repeatedly told me her daughter's implantation was a success because of the hard work she put into it. Carol also explained to me that "you have to do your homework, and you have to do a certain amount of work. . . . and I'm an academic. I have time to get on the phone and put my phone on hold and wait, wait, wait, wait. I cannot imagine what somebody who doesn't have their own private office, who's working, doesn't have time. I don't know how you do it. . . . I'm sure if you dealt with any insurance stuff, you know, you've got to get to the point pretty quickly where you know your system better than they know their system. I mean, I got to that point pretty quickly, so I could be on the phone and say, 'No, actually, I don't need preapproval for that, and you've got to give me the [insurance] code.'"

Therapeutic Labor Now Means Hope for the Future

The therapeutic and bureaucratic labor involved in working toward implantation requires resources and cultural capital but also becomes a way to fight against anxiety. Jane, like Julia and Paul, enthusiastically integrated multiple interventions and services into her home and Lucy's routine. The promises and imaginations of the future cast meaning onto the present tasks of mothering (Gentile 2011). And this looks different in the age of implantation. In the past, parents were "not led to an overview of the life-trajectory that is probable for their [deaf] child . . . [nor] led to imagine possible futures" (Lane et al. 1996, 36). Today, however, parents are led to believe what their child's future will be with an implant and encouraged to imagine it. Julia and Paul did so when they witnessed an adult CI user. Other parents did so when reading Strivright's claim of the unhindered or "absolutely unlimited" future of the child who receives intervention. Julia and Paul continue to do so when they reflect on all the work they do with Morgan. As a result, Julia said, "We have every reason to believe that he'll be intelligent, independent, and high-functioning."

This successful, implanted future, however, must be created in the present, where parents are told that the more they intervene and the earlier the child is implanted, the better. Yet the stages of implantation can be long, thus causing parents anxiety and making their experience of the time between diagnosis and implantation fraught. Many had wanted the CI as soon as possible and saw everything before implantation as "wasted time," without language. Some, like Jane, even began having panic attacks and experiencing other psychosomatic indicators of anxiety. Although, as I will detail later, there are also parents who are more hesitant and worried about commencing with surgery.

Another way to imagine a hearing future is envisioned through the Cochlear Americas marketing materials that are distributed to parents. They are printed on glossy magazine paper and include a promotional DVD titled "An Absolute Miracle!" I ordered my own copy from my contact at Cochlear. The main menu of the DVD has three sections: "Real Life Stories," "Learn about Cochlear Nucleus 5," and "See What the Experts Say." There are four patient stories, two pediatric and two adult. All of the featured families and patients are white and shown in middle-class, suburban settings.

One of the pediatric stories features a young girl who lost her hearing at age seven and was subsequently bilaterally implanted. She is interviewed, talking and listening, passing for hearing. Her story includes the "Hallmark moment" of initial stimulation (when the implant is first turned on) that Becky lamented in the previous chapter. There are also interviews with her teachers, who talk about how they no longer have to accommodate her in any way, that she is now just like all the other children. The second pediatric story features an infant, aged twenty months, who had been implanted five months prior. This story focuses on the parents' experiences, telling the story of "not passing" the NBHS, then following up with an ABR and being told that their daughter is profoundly deaf. The parents are relieved to learn of the availability of the CI and assured that everything will be OK.

The surgeon in this case is also interviewed; she says that the parents need to make the decision for their child right away to ensure that the child can perform well later. She adds that if they wait until the child is older, "They [will have] lost the battle." The parents in this story describe the surgery as having gone well, with no complications. They report that once the CI was turned on, their daughter started responding to environmental sounds. They talk about doing home speech therapy twice a week, even after surgery. The mother featured in the story emphasizes how crucial this is. This DVD is typical of the information parents consume before implant surgery, and during a child's candidacy appointment, audiologists ask parents if they have viewed the DVD.

In their post-diagnosis anxiety, sometimes parents want their child to be a candidate when they are not. "There are inappropriate implant candidates," Dr. Brown told me. Those are the "really hard evaluations" for Dr. Brown because it takes much more time to tell parents that their child is not an implant candidate when perhaps, on paper, he or she might look like one. In other words, the child might meet the requirements for type and degree of hearing loss. But, according to Dr. Brown, this does not take into account the fact that one must be deemed a candidate not only audiologically but also medically. Dr. Brown illustrated the extreme level of hope that parents associate with the CI through this story:

> For instance there was a child that we saw . . . that by audiological criteria they clearly met criteria to get an implant. And when I had done the imaging [MRI] on her, she had a normally formed cochlea, and yet the internal auditory canal that carries the hearing nerve from the cochlea to the brain had no auditory nerve running in it. . . . She has a cochlea. You can theoretically, surgically put the wire in, but you're not going to have any central connection. So, I advised against implantation. It was a hard thing to say, but . . . we all, as parents, care about our kids and we'll do anything for our child. . . . [They] got an implant for this child [at another clinic].

> And to this day, this kid has no response to the implant, is not getting anything, is nonverbal, and has no functional results of the implant. . . . The kid isn't really signing, isn't really communicating, has no oral language skills at all, and they've had an implant for a number of years.

It is not unusual for parents to have high expectations for the CI that are misaligned with its actual expected outcomes (Hyde, Punch, and Komesaroff 2010). When I asked the surgeon why parents often assume that the CI will always be successful, Dr. Brown answered, "The myth is, honestly, is human hope. That if you can read about it, you can be in that best possible category and you'll achieve it. The other thing is honestly, the marketing and literature for the CIs from the companies. They are such that they really lead you to believe that everyone can perform you know, fabulously with this, that there aren't grades of performance."

Determining Candidacy: The CI Team

The audiological and medical criteria for candidacy are what Dr. Brown refers to as "hard criteria." Audiological tests that determine the type and degree of hearing loss and amount of functional gain provided by hearing aids comprise one measure. Medical criteria include a full workup of the child's overall health and physiology, to make sure, for example, that the child has a cochlea (there are conditions where a child may be born without a cochlea or auditory nerve) and has no other precluding conditions. All of this is part of the standard process toward implantation candidacy and requires parents to schedule testing and imaging appointments, such as MRIs.

Clinic staff have long been monitoring parents' emotional states and compliance well before they undergo the transition into candidacy and prepare for surgery. They accomplish this in one way by holding monthly CI team meetings to share a variety of perspectives evaluating these soft criteria. When recommending a child for implantation, the CI team considers the parents' commitment, compliance, and understanding of the long-term requirements. They also talk about the EI service placement as well as educational placement if the child is older. All of these structures and clinical staff then work together to determine how well parents are being socialized into their new roles as effective parents of CI kids.

One morning before the CI team meeting, we were gathering the folders on the children who were to be discussed that day. I asked Sharon if parents knew that all of the clinic staff talked to each other about their cases. Often these meetings end up showcasing the different relationships that parents have with the audiologists, as compared to with the speech therapist, the social worker, or Dr. Brown. In fact, Sharon commented that when the group got together to discuss cases, often Dr. Brown was surprised by what the audiologists reported that parents would say to them, and in turn they were surprised by what they learned that patients said to Dr. Brown.

Although it was routine for the staff to talk to each other about the patients, formal CI team meetings were new at the center; I was able to sit in on the very first one. I was surprised by how many people attended. About ten of the clinic's staff reviewed a printed list of pediatric CI candidates. The staff in attendance included Dr. Brown, Annette, Monica, Sharon, Gretchen, and Sonya.

Sharon started going down the list of patients, and for each one, the group discussed a surprisingly wide range of issues that each child's family was dealing with. The team considered much more than just audiological and medical workups when discussing whether a patient should be implanted, and it was by sitting in on these meetings that I learned the complex dynamics and social criteria that contribute to determining candidacy. Discussions included which institutions the family was working with, which services they were receiving, whether the parents understood how to advocate for their children and form connections with institutions, and how well they were navigating social services systems. In other words, clinic staff were evaluating whether parents were engaged in concerted cultivation and achievement of that seamless overlap between home and intervention services. Lack of this achievement renders the child's candidacy questionable.

These meetings cover all spheres of a child's life—medical particulars, home, and school—all of which are considered "medical" information for the purposes of determining candidacy. When Dr. Brown brought up the first patient, the discussion switched momentarily to briefing the rest of the staff on the etiology of the child's hearing loss, the possible surgery and initial stimulation dates (initial stimulation occurs about four weeks after surgery), history of hearing aid usage, and patterns of hearing loss and functional gain. But after this, the audiologists went back to discussing the child's home environment. Monica stated that his parents are Spanish speaking, so that is a "double whammy." When I followed up with Annette about this comment later, she explained that exposure to spoken language with the CI is crucial, and that it must be consistent. Thus, there is not only an imperative for spoken language but specifically spoken English: English is spoken at school and in intervention services, and the team perceives speaking Spanish—or any other language besides English—as muddying that exposure process. The team decided to "wait and see" with this patient.

The next case brought up the question of whether the parents could demonstrate consistency and the ability to manage the technical aspects of the CI equipment. Because of the child's educational placement in an auditory-verbal classroom where spoken language was the focus, the staff were confident that use of the CI would be enforced during school hours. But they worried about whether the CI would be used at all times at home. Dr. Brown posed the question of whether they can know if the CI will be turned on during the weekends. Monica agreed; they would need the consistency. Monica added that they would also have to make sure someone checked the equipment regularly to ensure its programming was correct. But Gretchen felt the child should be implanted, saying that she had showed the child the audiogram to show functional gain with a CI compared to a hearing aid, and she felt he understood what good hearing meant and that a CI would be beneficial. Dr. Brown decided that there needed to be a big meeting with the family to discuss the many questions she still had regarding the patient. Dr. Brown was "uncomfortable" moving forward with the CI.

Assessing Emotions and "Accepting Reality"

Evaluating the social criteria for candidacy also includes consideration of parents' emotional state. During their encounters with parents, audiologists look for a process they often call "accepting the reality" of a child's hearing loss. Time and again, I heard audiologists at the center talking about the absolute necessity of parents "accepting their child's hearing loss." This is not resignation to deafness. Rather, the acceptance audiologists speak of is parents' accepting that they need to work to overcome it. For audiologists and other clinical staff, it is socially unacceptable *not* to choose the implant. In other words, what audiologists call "acceptance of reality" is code for whether parents had appropriately emotionally processed the child's diagnosis and accepted their recommendations for intervening audiologically to mitigate, alter, or circumvent that biological reality. Parental failure to move through those emotions and participate in biotechnological intervention—in this case, implantation—constitutes a failure to "accept reality." This is similar to Anspach's (1997) observations that physicians' habits of "psychologizing" parents making decisions about their child's medical care resulted in diminishing parental authority and undermining their status as rational decision makers. Decisions based on conflicting cultural norms are also implicated here, such that differences in families' cultural backgrounds are medicalized through a psychological discourse. This holds up the medicalized script of deafness as acultural and "objective."

Annette, the chief audiologist at the center, explained, "So you have those families where there's no question about it; how could you think I wouldn't do this? I can say it tends to be routine with normal hearing parents with deaf children." "Normal" parents decide to implant. This decision to implant is assumed as a given as parents participate in the interventions leading up to official candidacy evaluation. And this participation, Annette said, was "acceptance, really believing it." She also said that in noncompliant families, "denial is the big common denominator. . . . We've all had these families. It's probably a handful, but some stand out more than others."

When parents reject the idea of hearing aids, refuse to put them on their child, or do not actively participate in EI, they are often said to be in denial. Discussing a particular case with me where the child did not have any language, Annette became exasperated as she told me, "The fact that he's not talking, would you rather people think he's just stupid or mentally retarded? That's better?" Over and over again she said she just does not understand. "At what point can you keep this a secret? I don't get that. . . . I don't understand the thought process down the road when they do that. I tell them, you should put the hearing aid on him, but I can't go home with them and say do this. But at what point can you still keep this a secret?"

This kind of ongoing denial, she said, is rare but does happen. Sometimes it takes years, but she and Sonya "stay on" the families; eventually they may come around. She postulated some of the reasons for parents' denial, suspecting perhaps the stigma associated with hearing aids. "I think that's huge. That's huge! 'They're going to make fun of him.' What do you think is going through a parent's head when you tell them their kid is hearing impaired?" Similar to what many of the parents told me, she said what she imagined they were thinking: "They're going to pick on him at school. And their first thought is: Will my child be able to speak? Will I be able to communicate with him? And when he goes to school, he's going to be tortured."

Other times, Annette told me, parents have to become desperate in order to get "on board." She described one family to me she was excited had finally made an implant candidacy evaluation appointment. "He [patient] is starting to sign, and they can't communicate with him. So if they get a CI—well, now they're ready to get a CI like yesterday! They want to know how fast he'll start talking." This statement alone implies that (1) it is acceptable for parents not to communicate with their child in sign language, even if that is the language the child uses, (2) the child's use of sign constitutes a sense of urgent intervention to reverse that trend, (3) if the child were implanted, he would learn to use spoken language, and (4) by wanting the implant, the parents are "normal," compliant, and "accepting reality."

But what about those parents who do not accept reality? The audiologists refer to them as the "difficult moms." During a discussion with Monica about one of her "difficult mom" cases, she told me about a child who is almost five years old. The mother had had a "very difficult time just even accepting the hearing loss. It took her a long time before she would get to the point where she would allow us to fit her child with hearing aids because she was still going through acceptance." I tried to clarify what Monica meant by that, and she told me, "She couldn't really understand why sometimes [her daughter] would hear a sound and be able to turn, but she wasn't talking. That was probably because she did have some residual hearing, and so she could hear louder sounds. But she didn't have enough hearing to understand speech. That took an awful lot of parent education from not just myself, actually, from a lot of people, the social worker, other audiologists, Dr. Brown. It took a lot to get her to the point where she would accept the hearing aids." She surmised that it was difficult for this mother due to a mixture of denial and being unable emotionally to handle the situation. "She [patient] was actually evaluated for an implant three separate times. We've been wanting to implant that particular child for years already." Recently, the child had finally been fitted with hearing aids and "placed in an appropriate school program, getting a lot of services." But Monica and the rest of the staff felt that she had reached a plateau. Monica explained, "There's a plateau that you reach in terms of the clarity of the speech because she didn't have access to as much hearing as an implant could provide her with."

The problem, as Monica saw it, was that the mother did not understand that while the hearing aids were good, the implant was better: "She's coming along, but we knew she would do better with an implant. But it took again, like I said, we went through three [official implant evaluation appointments], before the mom would sign the [paper]." These evaluation appointments, which comprise a battery of hearing tests and discussions with the parents, took place over the course of a year and a half. "It took Mom a long time. The school was pushing, trying to get her to take the implant." In some cases, the difficult part for parents is the idea of surgery. She continued, "Sometimes that end of it is more of a difficult hump for people to get over because they are afraid of the surgery, afraid of the anesthesia. . . . But Mom is coming through, she's definitely making progress with the implant. She just had a tough time, but she's coming around." This frustrated Monica because she and so many other professionals working with the child had been pushing for the CI from the beginning. "But everybody is different. It depends on what their frame of mind is when they start the process; if they've accepted the hearing loss or if they haven't." But other parents "would be on our doorstep every day if we let them, so there's a huge, huge range."

Monica stressed that it was not just the audiologists who pushed for the CI. "Everybody on that child's team does the same thing, and the parents learn over time. School plays a very big role in that because the child is there at school the majority of the time and they offer a lot of parent education and working with the speech pathologist. The school is one issue; we on the other hand, we have a whole parent education meeting here actually. . . . The support group is run by Gretchen, and its role—well, it comes back to parents accepting the hearing loss."

For Annette, difficult moms were those who are "hard to get on board"; she emphasized to me that this is usually a cultural issue. Often, she can predict which families will present difficulties. She said, "Most of the families that don't implant are parents who are deaf themselves. . . . I can say, 'If you're interested, we can talk about the CIs,' but those are the families where [they say], 'If my child wants to get implanted, when they're old enough to ask for it, then we'll consider it.'" She does not expect them to engage in EI services, hearing aids, and implantation.

"Culture is big, but you know that already because you know the Deaf culture," she commented. But Deaf families are not the only ones she experiences as "difficult." While most "normal" parents do implant, she said, "the outliers are the ones who tend not to be American . . . they tend to be foreign-born parents who have more cultural issues that start to come into play." She recalled a family from Russia she had recently been trying to work with, surmising that "having an impaired child or putting or . . . having to put a hearing aid or a CI is so obvious." Once she had identified the baby as deaf, "They didn't even want to talk to me, and they left me a message telling me to leave them alone. . . . I have another Israeli family . . . they picked up and literally, I swear to you, they moved to Israel to get away from me." But she told me that the story ended well because the parents moved back, and after a long deliberation, they finally decided to get a CI for their daughter: "She's talking now, so everything's fine." In her experience, "culture is big," and those who do not implant tend to be "parents who are not native [English] speakers and parents who are not born here."

Audiologists deemed accepting reality as good parenting that resonates with white, normative, and middle-class ideals of developing one's child and emphasizes becoming an "independent" and "functioning" adult who can pass for hearing. Acceptance of reality for audiologists also means participating in culturally accepted ways of overcoming deafness. This echoes Rose's assertion that in a contemporary, biomedicalized society, "individuals are enjoined to think of themselves as actively sharing their life course through acts of choice in the name of a better future; 'biology' will not easily be accepted as fate or responded to with impotence" (2006, 26). While Rose speaks about individuals compelled to intervene upon their own bodies, what I have shown here is how this is classed, based on an ideology of ableism (or what Deaf studies would distinctly call audism), and intersects with expectations of scientific motherhood. Furthermore, I have also demonstrated how framing resistance to medicalization as "cultural" reinforces the category of medical as outside of, or in opposition to, culture; that is, our compulsion to intervene upon and normalize bodies is rational, while not doing so is constructed as irrational either through psychologizing discourse about the parents being in denial or through attributing it to having a different culture.

Preparing for Surgery: The Evaluation Appointment

Summer is prime time for children to get CIs. If they are school-aged, implanting them in the summer allows—without disrupting classroom attendance—time for the surgery and ensuing months of healing that follows before the device's initial activation. This scheduling also provides time for the child to acclimate to the flow of electrical current and the changing "maps" or frequency settings, a process I will describe in more detail in the next chapter. No matter the children's ages, summer is also better for surgeries since the warmer months mean fewer ear infections.

One particular day, a pediatric implant evaluation appointment (PIE) was on the schedule. PIE appointments are what Monica and Annette alluded to earlier, and they are long. That day, four-year-old Louis was coming in; Lisa, one of the other audiologists, and Monica and I discussed the case before he arrived. Lisa noted, "We did a sedated ABR a year ago and indicated that the loss was appropriate for hearing aids, but now we may recommend a CI—something must've changed." Monica nodded her head in agreement, "Right, we aren't sure if he's even a candidate for CI, so will do aided tests today and get a history from the mom. This appointment is PIE, and we won't know until we get the test results if he is [audiologically] a candidate for CI."

The first thing Lisa and Monica did was ask the patient's mother, Sylvia, if she had read all of the materials on the CI they had given her and watched the DVD from the CI company. She had. Then Lisa and Monica asked Sylvia how often she used the hearing aids with her son, what brands and settings she used, and if she had already had all the medical imaging done. She had; the CAT scans and MRIs had not shown anything that would preclude surgery. Lisa and Monica asked about Louis's current educational placement and overall history. Sylvia said he was set to transition into regular school in the fall and noted that for some time he had seemed to have residual hearing and auditory memory. Louis was receiving speech/language therapy.

In PIE appointments, the clinic determines candidacy by seeing how much benefit hearing aids are providing. In these appointments, we used evaluation rooms that had an enclosed sound booth; one audiologist sat on one side, while the patient, the mother, and the second audiologist sat on the other side. A small window joined the two rooms; everyone on both sides could see each other. Lisa and I went into the booth with all the testing equipment—panels of machines with various knobs and levers—while Monica sat with Sylvia and Louis on the other side of the window.

Lisa proceeded to present tones at varying frequencies and decibels, while Monica determined his response. They measured the difference that the hearing aids made in both ears in order to determine the functional gain. Meanwhile, Sylvia filled out a questionnaire that asked questions about how Louis functions with the aids and about the behaviors she noticed. The answers were scaled; this information—along with results from the functional gain testing and Gretchen's speech/language development reports—were considered in the audiological criteria for implantation.

During a break in the testing, Monica came back to the booth to look at some of the results with Lisa. I asked Monica if she thought that he should get an implant; she said, "Yes, based on the amount of benefit he's getting from aids." We then moved from the testing booths into one of the appointment rooms with observational mirrors. I sat behind the mirror and watched as Monica sat down with Sylvia and said, "Let's go over things. If we can raise the level of amplification, which is maxed out right now with hearing aids, we can give him more access. He'll do better with a CI. It will also help with his behavior; he has to work really hard to hear, and with a CI he won't have to work so hard." Monica emphasized that they only had partial information at the moment: Sylvia would need to bring him back in for additional speech perception testing. But she assured Sylvia, "Yes, I think he can benefit from the CI, but it's not just my decision." Monica described the CI to Sylvia, showing her one that they keep on hand as an example.

"Audiologically, This Patient Is a CI Candidate."

After this appointment, I looked through some of the medical records of CI patients that are part of my study. As I did so, I saw the phrase "Audiologically, this patient is a CI candidate." Although I had begun to understand there was more to being a candidate than just audiological evaluation (for example, I also knew that medical clearance had to be determined through MRIs and other imaging techniques, like CAT scans), I still wanted to know more about the social criteria that had come up during the CI team meetings.

I asked Shelly, another audiologist at the center, if she had a few minutes to explain candidacy to me. She replied, "You qualify audiologically if you have the audiogram for an implant, the horrible speech, and you're medically intact to have an implant. But if you don't have the support system at home, or you don't keep hearing aids on, then you won't succeed with it. Then that's where it gets to: Are you a candidate or are you not?"

She explained further, saying that there are social factors to consider. Unlike the "double whammy" comment I had heard in the CI team meeting, however, Shelley told me this story that illustrated her belief that speaking another language should not disqualify a child from candidacy:

> In New York, we have so many cultures. I remember being in CI class in my online program for my doctorate. I had to do a group project, and I was placed with people from Kansas, Wyoming, everywhere but New York. We were given cases, and we had to pick who was a candidate for an implant. There was a little

> Korean family, and the mother only spoke Korean, and the father was in denial about the implant. Dad is in denial. Mom is ready to go forward with the implant. I said, yes. I got yelled at by everyone in my group. "They don't speak English!" I said, "You're from Kansas. No one speaks English when you come to New York, you learn that's how it is." They wouldn't listen to me. We get the project back. We got that question wrong. I was right, he was a candidate. The teacher said just because he doesn't speak English doesn't mean he can't get an implant. And what happened with this case was, she said, he got an implant, Mom took classes and learned English, and Dad got like therapy to realize his kid had a hearing loss.

Family context clearly plays a role in decision making about candidacy, but that role is not always clear-cut, and as in Shelly's story, it is not nationally uniform or agreed upon either. "It's a whole big battery of pieces if you think about it," she told me. As I looked further into this, I found that there are currently no national guidelines for determining candidacy, and this is a cause of great concern for professionals in the field (Sorkin 2013). "With the lack of evidence-based standard of care, patients are seen as-needed by the audiologists and the schools or early intervention systems are responsible for developing and implementing the aural (re)habilitation program" (Bradham, Snell, and Haynes 2009, 32).[5] Thus, CI candidacy assessment varies from center to center, but the FDA guidelines call for medical evaluation and "audiological testing at two points in time to demonstrate degree of hearing loss and (lack of) auditory development" (Bradham, Snell, and Haynes 2009, 35). In other words, the audiological criteria for a CI include documenting a sustained period of time where there is a *lack of language* and/or the *lack of access to sound*. The American Speech-Language-Hearing Association (ASHA), however, additionally recommends asking the following: "Do the necessary supports exist in the individual's psychological, family, educational, and rehabilitative situation to keep a cochlear implant working and integrate it into the patient's life? If not, can they be developed?"

The Pre-Op Appointment

One morning in July, Jim and Tina sat in the waiting room with their daughter, Amanda. She was twelve months old. This was the final appointment with Dr. Brown before Amanda's surgery; hence, it was less remarkable that both parents were present. Jim and Tina are a white, middle-class couple in their thirties. When the ENT receptionist finally called them back, I helped them gather their things, and then we all squeezed into Dr. Brown's small exam room.

Dr. Brown greeted them, saying that this is it, the last time they will see each other before the surgery. They had pushed the surgery back one month to wait for the newest version of Cochlear Americas' Nucleus Freedom, which was sitting in a box on the counter. When Dr. Brown pulled it out, we saw it was white and included upgraded software and a remote control. It looked like an iPod.

Dr. Brown explained again the reason for waiting that extra month, saying, "The difference between a new internal [piece], and the reason it's significant in someone her age, is do you see the height of the two implants?" Dr. Brown held them both out, then continued, "This one [the newest one] is a lot flatter. One of the problems with the Freedom used to be that it used to stick up a fair amount from the scalp. This [new one] has a much lower profile. What happens is it lies much closer to the skull, and it doesn't protrude as much."

Dr. Brown gave the new internal piece to Jim and Tina to examine. "If you want to feel the two, there's a big difference, because in an infant, you actually have to drill all the way through the skull down to the dura, which is the linings of the brain." But the newer model is different. The old model "actually presses into it because of the profile, whereas this newer one, what's protruding from underneath is a lot less," Dr. Brown explained.

Tina asked Dr. Brown if that was the only difference between the two models: "There's no real upgrade in terms of the electronics themselves?" From a surgical standpoint, Dr. Brown told them, there was not a significant difference. "Drilling a well is drilling a well, and I'm just going to change its shape a little." And regardless, all of the rest of the surgery's procedure was identical. "It's the same in terms of the mastoidectomy, going over the facial nerve into the middle ear opening, the inner ear, that's all the same. Opening the cochlea and putting the implant in is the same; all of that is unchanged."

Jim and Tina looked at each other and nodded. They were calm but inquisitive. Dr. Brown asked them whether Amanda had had any recent ear infections. She had not. Dr. Brown lifted Amanda up onto the examination table, exclaiming, "Oh, you're a wiggly worm-and-a-half!" Amanda started crying. Her hearing aids were on, and Dr. Brown wanted to look in her ears. "Oh! I'm sorry," Tina said, "I'll take your hearing aids out. How can we look in there, if those hearing aids are in?" Dr. Brown looked into Amanda's ears and asked if they had made their appointment for the initial stimulation. They had. Dr. Brown complimented Amanda in a high voice, "You're so good!"

When Dr. Brown was done examining Amanda, Tina put Amanda on the floor to play while the three adults reviewed the surgery details. "Basically, the incision is pretty much the same. It goes from here to about here," Dr. Brown ran a finger behind the ear, motioning from the top to bottom. And continued, "She'll wake up with a head wrap dressing on her head. If she feels great, she could go home the same day. If not, certainly, most people I keep twenty-four hours. Parents that want to go home the same day, that's fine." Dr. Brown explained that for the next three days, the incision could get wet, but there could be no scrubbing: "It's critical to keep her fingernails as short as possible. You don't want her scratching at that wound. If she scratches that, she could introduce a local infection. That's one of the biggest issues in young kids." Jim and Tina nodded.

During the surgery, the implant is tested to make sure it is working, and it is examined by X-ray. "The X-ray tells me the position, that I like it, that's in the cochlea, that I don't have concerns about where it is," Dr. Brown said, before reviewing all the risks associated with the surgery. These included the standard risks with anesthesia, the possibility of facial nerve damage that could result in paralysis, and a higher rate of contracting meningitis. Parents did worry about anesthesia, but none of the parents I interviewed particularly emphasized this fear. Complications like meningitis are rare and prophylactically treated by vaccinating children before surgery. Similarly, surgeons avoid facial nerve damage by using a monitor during surgery. "I will use a monitor. It tells me where the nerve is and how to work around it safely. I've never injured a facial nerve," Dr. Brown assured Jim and Tina. The parents nodded, looked at each other, and said OK.

Surgery Day

I met Jane and her husband at the hospital around seven on an August morning. It was Lucy's surgery day. Jane and I had talked the day before; I had called her after I had seen Lucy's surgery scheduled on the computer. An hour after our phone call, Jane said, the clinic had called her to give the instructions regarding not letting Lucy eat or drink anything. Everyone in the center was talking about Lucy's upcoming surgery because it had been such a long time coming. In the waiting room that morning, I met Jane after she and her husband had taken Lucy back to the OR. "They are drilling a hole. They are touching the brain stem. It is minor brain surgery. It is *scary*. I know Dr. Brown has never [harmed the facial nerve], but her face could be paralyzed. What if we did this and it wasn't the best thing to do? What if the electrodes don't take?" Jane worried.

She sank down on the couch as a TV blared out CNN. I had brought some snacks and offered to get coffee while we waited. "We got here at 5:45 this morning!" she told me. She looked extraordinarily tired. "At least you know how my anxiety manifests," she said. She looked down at her hands and touched her own fingers. I asked her if she had eaten anything, but she had not. Her husband had forgotten the sandwich that she had wanted. Nothing else sounded good to her. "I love my husband, but I think he was born with a defect when it comes to emotional support," she said. Meanwhile her teenage son was at home with the other two young children. She paced. "My head is there [at home], but my head is here too."

While we waited, Jane told me that Dr. Brown had had to call Cochlear Americas that morning because Lucy was having the Nucleus 5 internal implanted. "But I ordered the old external, nothing internal has changed, but there is no FDA approval on the new battery," she told me. I was puzzled. Why did that matter? "If you get that one [the newer model], you are spending zillions of dollars on batteries. . . . The CI uses three batteries. They have to approve every part of it, the coil, every piece. If you buy a package of batteries for $195, it will only last a few weeks. That's why I went with the old unit that has FDA approval on rechargeable batteries; they're much cheaper and have a one-year warranty," Jane said. She was already in the know and navigating the equipment—one more step in her mastery of all the complicated pieces of implantation. A couple of hours later we were called to the recovery room; Lucy was successfully out of surgery.

Concerted Cultivation, Language, and Good Parenting

While these audiologists never explicitly discussed class status, their suggestion that "normal" parents implant indicates tacit approval of particular parenting styles. In this case, a parenting style that aligns with concerted cultivation is one that shares the same value system as the institutions responsible for serving deaf children, which may explain why my sample of highly compliant families is overwhelmingly white and middle class. Suggesting that other cultures, class statuses, or value systems do not align with the high-intervention/concerted cultivation values of the clinic offers a more complex sociological picture that sharply contrasts with audiologists' assumptions that those who do not opt for implantation simply do not "accept reality."

Without necessarily situating Lareau's work within the context of parenting a child with a hearing loss or even of deaf education, we can see that her findings turn out to be quite relevant in explaining the class or cultural background disparities in implantation. For example, she finds that language is an axis of differentiation across social classes. Middle-class families have a *relationship* with language: "they enjoy words for their own sake, ascribing an intrinsic pleasure to them. . . . parents use language as the key mechanism of discipline" (Lareau 2003, 107). By contrast, she finds, poor or working-class families use language for more functional purposes.

But what happens to these kinds of language usage patterns when they are disrupted by hearing loss? These very patterns and relationships with language that appear in middle-class families in Lareau's study also frame the willingness of the parents in my study to engage in sustained therapeutic labor for the sake of their children's futures. In middle-class families, language skills are seen as part and parcel of middle-class life; that is, using language to your advantage, along with skills in negotiations, "is an important class-based advantage" (Lareau 2003, 111).

Although sign language is linguistically equal in that it is as legitimate a language as any other, its status as a minority language misaligns it against broader patterns of language use in middle-class homes. Thus, a high-intervention parenting style that could result in making the child's deafness invisible through implantation and enough auditory training becomes what parents envision for their child's future. Imagining a child who is not implanted and does not use spoken language (and instead uses sign) is incompatible with such middle-class values. In the following chapter, I turn to how acceptance of a "new reality" after implantation is centered around the brain.

4

THE NEURAL PROJECT

The Role of the Brain

The human brain does not discriminate between the hands and the tongue. People discriminate, but not our biological human brain.

• Laura-Ann Petitto

Neurons that fire together wire together.

• Hebb's Law

The brain is the focus of all stages of implantation. Implantation professionals assert that "identification of newborn hearing loss should be considered a neurodevelopmental emergency" (Flexer 2014). This focus on the brain is occurring in a broader social context wherein neuroscientific explanations are particularly popular as a field of research. In 2013, President Obama unveiled the Brain Research through Advancing Innovative Neurotechnologies (BRAIN) Initiative. BRAIN aims to map every neuron of the human brain. As a result, neuroscience is also an increasingly powerful trope of explanation in society and a subject of sociological inquiry (Pickersgill 2011; Pickersgill, Cunningham-Burley, and Martin 2011; Vidal 2009; Vrecko 2010; Rose and Abi-Rached 2013).

The neuroscientific explanation of deafness, however, emerges in relation to prosthesis; the body and the technology are bundled and inscribed with social relations together (Jain 1999). In the previous chapters, I showed that deafness is not inherently or naturally a problem but coconstructed as such through a medical script of deafness intertwined with the technology of the CI. In this chapter I focus on how the meaning of deafness is constructed specifically in relation to the brain. As a result, the definition of deafness is redefined from a sensory (hearing) loss to a neurological (processing) problem.

As deafness is framed as a neurological problem, the CI is simultaneously framed as the answer to gaining access to the neuronal structures in the brain so that they may be shaped. The potential to shape one's brain, the emergence of the CI, and this redefinition of deafness ultimately shift responsibility *from the device to the individual;* that is, the CI provides access to the brain, while the "real" treatment emerges as something else, namely, long-term therapeutic endeavors focused on neurological training required of mothers. And these endeavors begin from the moment of identification; they frame interventions to prepare the child's brain for the CI later as well as shape the interventions once surgery is complete.

This capacity to shape neuronal structures and neural pathways that is prioritized in implantation is known as neuroplasticity. The idea of plasticity fits seamlessly with the therapeutic imperative and concerted cultivation parenting style. It triggers an ensuing responsibility on the part of mothers to train their child's brain and to ensure proper development of the "right" synaptic connections for spoken language. But it is important to note that only certain narratives of plasticity are typically deployed by professionals in relation to CIs. Plasticity is not just about being able to change neural pathways or synaptic connections; it also refers to the brain's capacity to adapt, respond, and change as needed. The narrative of plasticity in implantation is that the CI allows you to shape a deaf child's brain into one that has capacities similar to a hearing person's; that is, the CI allows you access to the brain in order to cultivate auditory pathways for spoken language despite the fact that the child is deaf. This narrative of plasticity does not, however, allow for the possibility of auditory language and visual language pathways to be created simultaneously, even though the brain is biologically able to do so. As Fjord also found in her fieldwork, "US professionals in pediatric implants (not neurologists) [saw] signed languages 'hard-wiring' the brain *against* the capability to learn spoken languages" (2010, 87). As a result, it is common to suggest that parents prevent their child from being exposed to sign language. And yet, "contrary [to] frequent assumptions . . . there is no published evidence that sign language interferes with spoken language" (Knoors and Marschark 2012, 294).

The mantra in audiological practice that sign language is an impediment to speech acquisition has been around for some time, although the exact origins of this belief are unknown. Douglas Baynton's historical account of sign language in America found that as early as 1900, those who favored teaching deaf children to speak referred to instilling "habits of speech in the young," lest the "fatal habit" of signing take over (1998, 64). Today, however, the cultural preference for spoken language is articulated through the authority of neurological discourse, as Fjord's (2010; 2001) fieldwork in the United States in the mid- to late 1990s documents, and as the data described here, which were gathered in 2009 and 2010, also show. Thus, this has been and continues to be a persistent ideology in audiological practice.

All of this results in the displacement of failure from the device onto mothers' labor *and* a greater demand for adherence to auditory training and spoken language–only protocols characteristic of the medicalized script of deafness. Pitts-Taylor discussed this responsibility for "neuronal fitness" in the contemporary neoliberal, biopolitical context: "The ideal subject constructed here should see herself in biomedical terms and should relate to her body at the molecular levels. . . . brain potentiality represents a competitive field in which one's willingness to let go of sameness, to constantly adapt, and to embrace a lifelong regimen of work on the self (and on one's children) are the keys to individual success" (2010, 644). Thus, the extension of technological capabilities—these kinds of ever-deeper possibilities of shaping and developing minds and synapses—comes with an expansion of obligations for mothers. These capabilities are *both* obligation (ongoing, often invisible therapeutic labor) and possibility (for some of the implanted children I observed *do* appear to speak and listen like hearing children). In this chapter, I look back on the steps leading up to surgery through the lens of neuropolitics, make explicit the role of the brain in long-term habilitation efforts, and show how all of this cannot be untethered from the politics of language.

Connolly defines neuropolitics as "the politics through which cultural life mixes into the composition of body/brain processes and vice versa" (2002, 14). Too often, "cultural theorists reduce body politics to studies of how the body is *represented* in cultural politics. They do not appreciate the *compositional* dimension of body-brain-culture relays" (Connolly 2002, 14, original emphasis). For this reason, I approach the parent-child interactions in implantation as "body-brain culture relays" that take place in a cultural realm—that is, the clinic is a site of *cultural production*—that constructs deafness in neural terms. These relays occur out of a desire for and belief in the mutability and transformability of the brain and directly attend to the compositional dimension of perception; that is, perception is *made* through specific techniques of caregiving deployed through CIs.

The brain is integral to the very explanation of how sound, hearing, and the CI work. When a sound is made, sound waves travel through the air. The shape of the ear works as a cone that funnels these sound waves into the middle ear. The middle ear begins with the tympanic membrane, or eardrum; the eardrum is hit by these waves and vibrates the tiny bones behind it. After the outer ear picks up acoustic sound waves and sends them through the middle ear, they enter the cochlea, a tiny, snail-shaped organ in the inner ear. This organ, made up of spiraled canals, is lined with tens of thousands of tiny hairs called "hair cells." Here, acoustic waves convert to electrical signals, which are sent to the auditory cortex of the brain. The brain processes that information through synaptic connections, creating and depending on neural pathways to give the sounds meaning.

The CI takes over the job of the hair cells. The external component of the CI looks a bit like a regular hearing aid; a tiny microphone and processor—called a "behind-the-ear" unit, or BTE—sit on the back of the ear. The BTE has a small wire that connects to a round, magnetized transmitter coil. This coil sits on top of the skin, hovering over the internal component that has been surgically implanted behind the ear. The microphone on the BTE picks up sound waves and sends them to its microprocessor. Then, this signal is transmitted through the skin via the coil to the internal component. The internal component, surgically implanted by drilling a well into the skull and nestling the implant in place, processes this information; it is connected to a set of silicon-covered electrodes threaded into the circular canals of the cochlea. The internal piece sends a digital, electrical signal—a band of ones and zeros—to this electrode array, firing an electrical signal directly to the auditory cortex in the brain.

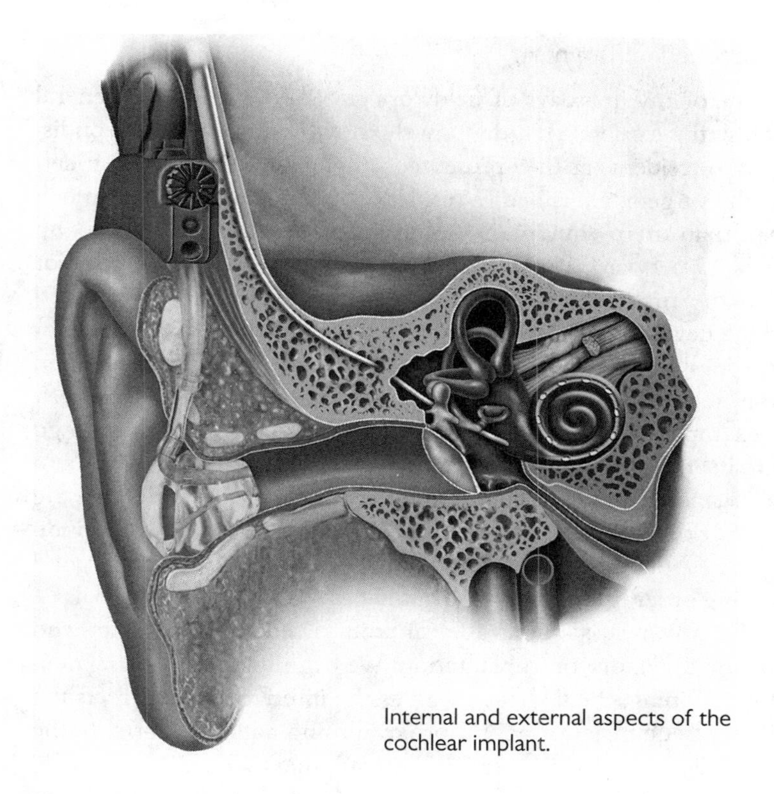

Internal and external aspects of the cochlear implant.

To be sure, there were methods of treating deafness before, but the technology of the CI has changed them. The neurological language that describes how the CI works refocuses the site to be worked upon to overcome deafness. The discourse is one of codes and circuitry that conjures the brain. The development of a neuroprosthetic treatment has opened up the realm of the brain, suggesting that the brain is the location of difference and that, therefore, if worked upon hard enough, that difference can be made invisible.[1] Although the neurological aspect of implantation has thus far been implicit, I turn now to how this focus on the brain has explicitly organized the social relations I observed from the very beginning, and how it has implicated the brain in ongoing postsurgery habilitation efforts. As Jane once told me when I asked her about life after Lucy's surgery, "This is where the rubber meets the road."

First Stop: The Brain

On one of my first days of fieldwork at NYG, I sat in on a general ENT staff meeting. That day, a well-known visiting CI surgeon listened to residents as they presented tough cases. During this meeting, the surgeon explained to residents that implantation was much more than mere equipment; some synchrony had to happen "up north." Afterward, I met with Annette, the chief audiologist, for more information about this comment. She explained that implanting the device is not the treatment but rather a necessary step that precedes the training of the brain—and that this training of the brain is the "real" treatment. This is why the brain—through an ABR test, for example—is often the first object of testing for infants who go on to be implanted. It is also why complete workups with MRIs and CAT scans are required before implant surgery, to be sure there are no other organic neurological impediments that would contraindicate CI candidacy and success. To train a brain, it must be in good working order; the "circuitry" should be in place and functional.

The ABR test is a neurological point of entry for patients who start the trajectory of implantation. We begin with measuring neural functioning; the CI lies in wait as the interface that corrects the missing frequencies, a neural programming patch of sorts. If the ABR indicates a missing signal, the equipment (the CI) is immediately implicated; it is ready and waiting to be installed in order to allow access to the brain and supply the missing audio signals. The Cochlear Americas promotional video features an audiologist who says of a patient, "She had all the wiring in place"—implying that the patient just needs the CI to use it. This patient's mother follows up by explaining that her daughter's brain learned how to hear once the implants were turned on.

Neurological discourse is present from the beginning and continues throughout every step of the implantation process, framing each task at hand for parents. For example, when discussing a child who had hearing aids but was soon to get a CI, one of the audiologists explained to the parent that "he has damage, so information isn't getting sent to brain . . . but the CI is a sophisticated computer that holds four programs, or 'maps.' The computer changes these programs to code, which is then sent through the coil, which communicates with the internal piece, and this fires and sends information to the nerve. . . . We send you home with four maps so that you gradually turn it up more and more. You have to get used to the electrical stimulation, so you may not be able to hear speech in beginning. He will have a lot of appointments at first, as we bring up the CI. We do it a lot; it will be fine."

At the same time, however, I observed that the CI was downplayed as merely access equipment even as it is promoted as an answer to particular types of deafness. "Failure" of the implant is understood only in terms of the functionality of the device itself and considered largely avoidable. Framing the CI as merely an access device neutralizes the notion of its potential failure. During Jim and Tina's final clinical appointment with Dr. Brown before surgery (described in the previous chapter), Jim asked, "So what's the failure rate?" Dr. Brown responded by asking, "You're talking about immediate failure versus later on failure?" The father clarified that he was asking about both. The surgeon explained:

> So immediate failures, there really isn't, OK, because if you put in the implant and we're not getting responses off the electrodes and we're not getting the—you know, the X-ray doesn't look right, I'm going to put in my backup implant. I always have more than one around [in the operating room] for the instant, you know, maybe if it's an off-the-shelf failure. Maybe that implant didn't work. That's why we test. That's why we want to know before we close up [the surgical site] and say, you know what, this implant is a good implant. That's the whole point of doing the NRT.[2]

For Dr. Brown, whether the implant works is not a question of long-term results (those were questions of the brain and of the labor undertaken to train it, not the CI) but rather a question of whether the device is operational. The first support group I attended, where I met Nancy, the "old-timer," she put it quite succinctly: "I don't know that there are many children that the CI did not work for, and often it's the case that the implant works fine, but it's the parents who don't do what's needed." So the CI is constructed as only a conduit to the brain that provides an auditory signal through technical specifications. It delivers a simulation of sound; it makes one technically able to hear. The rest of the work lies at the other end of the electrodes' signals: the brain.

After surgery, a child is typically sent home the same day, or at the latest, the following day, with the incision left to heal for about four weeks. Once the incision is healed, the parents bring the child back to the clinic for the activation—or "initial stimulation"—appointment. This is referred to as the "initial stim" date. At this appointment, the external components of the CI—known as the behind-the-ear, or BTE, unit, which has a wire with a small magnetic coil at the end of it—are connected to the internal parts. The external coil, which is also a transmitter and passes the signal across the skin, is placed on top of the device implanted and stays attached because it is magnetized. The magnetized coil acts as a radio transmitter and is attached to the BTE by a wire.

When the implant is turned on, sound waves arrive at the BTE's microprocessor, which takes this information and translates it into a digitized signal made up of ones and zeros. This information then travels through the magnet and into the electrodes snaking through the inner ear. This digital signal then fires up from the electrodes to the auditory cortex of the brain. The number of electrodes varies, depending on the model of the CI; during my research they had twenty-four. In a hearing person, the hair cells, which number in the tens of thousands, communicate electrical signals to the brain. By contrast, an implanted person gets information from, at the most, these twenty-four channels. The sound that arrives at the brain does not arrive as a discernable signal, however. To a brain that has not had access to sound before, it is foreign input, data without a codebook. Later in this chapter, I will talk more about the ways this signal travels through the brain.

Mapping

The precise configuration of this digital information (e.g., which frequencies and at which levels are extracted from space and transmitted through the CI) vary from patient to patient. They are dictated by the microprocessor's program settings, which are called "maps";

CIs hold four different possible maps, often referred to as P1, P2, P3, and P4. The CI user can use different maps in different settings, such as one for loud environments, one for one-on-one conversations, and so on. Maps are tailored to each child's optimal usage in an appointment dedicated to mapping.

Making sure that the device is properly mapped is one important aspect of implantation. It is also one of the ways that parents are judged in terms of how "on top of things" they are. All patients must have a certain level of hearing loss unaided to be considered severely or profoundly deaf, but each child's loss is different. Decibel levels and frequency thresholds are unique to the individual. The CI, if configured properly, is tailored to mitigate the difference between a "normal" audiogram and an atypical one, and no one child may have the same audiogram or comfort levels as another. Mapping involves setting thresholds, known as T-levels, which are the softest sounds the CI user can hear. The other component of mapping involves adjusting the comfort levels, or C-levels, which helps the CI user avoid overstimulation or even discomfort.

Mapping is the most important aspect of maintaining optimal CI functionality, but children cannot provide the user information needed, due to the complex nature of the concepts—most infants undergoing mapping do not have any means of communication (Mertes and Chinnici 2011). Thus, the pediatric population requires a bit of guesswork. Adult patients' subjective descriptions of T- and C-levels have helped audiologists develop methods of programming maps for children. But it is still a process of trial and error that can take months and years, one that requires constant tinkering.

While maps must be carefully calibrated or configured in the audiologist's office, it is mothers who have to follow the prescribed mapping plans at home, as well as help determine if they are working or accurate. This can be quite difficult. The process of mapping begins at the initial stim appointment, with a series of initial, low-level signals designed to slowly acclimate the child to auditory information and electrical stimulation. The initial mapping sessions are close together. One afternoon at Jane's house, she opened up her weekly planner for me. She flung it down on the kitchen counter while we talked about Lucy's acclimation to the implant. She had recently emerged from the thick of the initial mapping sessions. Each day in her planner was covered in letters and numbers; she had used different colored pens to differentiate the numbers scrawled on each day.

"Mapping is a lot less now than it was in the beginning," Jane explained. In fact, when Lucy was first "turned on," they were at the audiologist about once a week: "Oh god, she was flying through the programs!" Each time they went, the program changed; Jane had to write it down, track Lucy's response, and know when to "bump" to the next one, gradually increasing Lucy's levels of stimulation. Jane said, "You have to write the programs down every time you bump her up. See? [She pointed at list of numbers.] P3, P1, yeah, I had to write it down." She let out a heavy sigh. She had been carrying around the calendar with her for weeks and brought it with her to every appointment: "I have to bring in all this [paperwork, calendar]. In the beginning I had to know because the appointments were so close together. So like, OK, today is P1, tomorrow is P2; she already moved through all programs! There are four: P1, P2, P3, and P4. You have to be through to P4 before you come back for the next mapping. And [she started talking faster, not quite catching her breath] we moved through it, and I haven't had to write anything down because she's been on P4 and we go for a mapping on the fifteenth at ten a.m."

While audiologists focus on getting the maps technically right, parents have to constantly manage the programs and determine if they are effective and comfortable for the child. This is stressful and anxiety producing and usually dependent on intimate knowledge and observations in the home or other daily environments. For example, perhaps the child prefers P2 in certain contexts, while seeming to do better with P3 in others. Each child's level of tolerance is not uniform. (*Tolerance* is usually the word audiologists and parents use to describe moments when they realize the CI is on too high or the C-level needs to be adjusted—the child winces or cries.)

To complicate matters, depending on the age and school placement of the child, there are always new situations or spaces that indicate CI reconfiguration.[3] For example, if there is an FM system at the child's school, the map will need to be adjusted for the frequency variation in this kind of input. Jacks that wirelessly receive a signal directly from the FM system can be attached to the BTE so that the input from the FM system can be sent directly to a child's CI. Every space is potentially a new map reconfiguration, every space a potential source of anxiety.

Reading and interpreting the child's reactions over time, often through means that are much harder to delineate, are how parents learn the child's needs and preferences. Usually, parents told me, problems would manifest in other behaviors. Perhaps when the programs are changed, the child struggles with bed-wetting, has problems toileting during the day, throws tantrums, shows a lack of attention, or cries. This is a frustrating guessing game, and parents often told me their concerns about such behaviors. Since the CI is a one-way device, it does not respond to the brain's modulations or needs: it can only fire a preprogrammed signal (Chorost 2011).

Mothers often pick up on the need for mapping changes by assessing their children's behaviors and emotions. It can be especially hard at the beginning of CI use, just after the initial stim. Becky described what happened when her daughter Amy had her implant turned on: "It was a whole other story because Amy was terrified. She did not want to wear the implant. . . . She rejected it. She ripped it off her head. As much as I put it on, she was pulling it off. It was a battle for a good few months—a good few months. . . . [School was] enforcing it. I was enforcing it at home. As hard it was. She had tantrums and everything. But then she started to tolerate it. It got a little bit better and a little bit better."

A few months later, after Lucy had been implanted and had been in her new auditory/oral school program for about a month, Jane told me she was concerned: "Lucy is almost five years old, and she just started shitting her pants. That's something psychological." In support groups or on online forums, parents often swap stories of what works for their child, or patterns of behavior they notice, and how to tweak the device's settings. Jane was beside herself; although her level of anxiety had subsided so that her fingers were no longer tingling as she talked about it, she was worried. But other days Jane was exuberant; in between displaying these behavioral issues, Lucy impressed her teachers with her speech articulation. Meanwhile, Jane still frequented these parent support groups and online forums, which often focused on tips for managing or configuring the device, as well as dealing with other behaviors.[4]

Auditory Training as Neural Self-Governance

The task of auditory training—both before and after surgery—remains the primary therapeutic intervention. While parents have to monitor the technical aspects of mapping, an implanted child's neurons must also learn to perceive and decode the signal as not just sound but meaningful information. Audiologists and educators often refer to auditory training, or the process of creating these meaningful synaptic connections, as creating "the right neural pathways for language." Chorost, an adult CI recipient, writes that "training makes an *enormous* difference. Research in neurobiology is showing that the right kind of training has directly measurable effects on people's brains" (Chorost 2006, 173, original emphasis). The brain learns and creates meaning by connecting neurons together. Thus, the CI is held up as a method of retrieval, a way to cultivate the latent hearing brain inside.

Mothers are expected not only to maintain proper maps on the device but also to consistently expose their children to the auditory signal by keeping the CI on at all times. The idea behind auditory training is that consistent presentation of the auditory signal paired with particular therapeutic techniques will create the "right" neural pathways for language in the child's brain. But how are these neural pathways constructed? Hebb's Law states that "neurons that fire together, wire together." The brain is composed of billions of neurons. The neuron's job, with its many branches that reach out and communicate across synapses to other neurons, is "to accept spikes of electricity—called action potentials—from its dendrites. Depending on that input, it 'decides' whether to send an action potential of its own to other neurons. . . . A neuron is basically a tiny deciding machine" (Chorost 2011, 75). The goal of implantation is to send a signal to the neurons in the auditory nerve, and then (through long-term auditory training) to manage and control the decisions that these neurons make. In other words, the goal is to intervene so that neurons make "good decisions" that create connections between certain neurons:

> [Neurons] constantly send out new projections that touch other neurons. This happens most promiscuously in the infant brain, but it happens in brains of all ages.

> At the initial contact, the connection—the synapse—is quite weak; the dendrite can easily detach and hunt for another neuron. But if signals pass through the synapse, the connection becomes a little bit stronger. Pass signals through many times, and the connection becomes very strong and ultimately permanent. (Chorost 2011, 40)

For anyone familiar with neuroscience, this is not novel. For anyone familiar with implantation, this is why auditory training, which works upon these synaptic connections, is said to be of crucial importance. "Neuroscientists generally agree that the brain is an active co-creator of perception, not merely its recipient" (Chorost 2011, 43). While biological factors affect how neurons behave, the direction of neurons' decisions during auditory training is *purposeful and social*. Parents are taught to undertake this task through highly orchestrated and structured exercises.

EI as Neural Anticipation and Discipline

EI techniques used in the home are structured by neural anticipation and built around the eventual arrival of an auditory signal. In other words, EI is a conduit for instilling neural discipline. And it is a coconstructive effort happening through the body-brain-culture relays between the parent and child that are integrated into daily life. The caregiving techniques that mothers incorporate through EI and auditory training can be understood through two critical characteristics of neuropolitics: relational techniques of the self and micropolitics (Connolly 2002). Relational techniques of the self are "choreographed mixtures of word, gesture, image, sound, rhythm, smell, and touch that help to define the sensibility in which your perception, thinking, identity, beliefs, and judgment are set" (Connolly 2002, 37). Micropolitics refers to how these techniques are "organized and deployed collectively by professional associations, mass media . . . charitable organizations, commercial advertising, child rearing" (Connolly 2002, 38). Indeed, the micropolitical aspects of these relational techniques of self—such as child rearing—illustrate the "critical functions the institution performs in organizing attachments" (Connolly 2002, 38).

For example, when I sat with Julia and Paul, Morgan's parents, Julia explained how Marianne, their EI therapist, worked with Morgan "before he could even hear, just to build the patterns." When explaining these patterns to me, she reached over the couch and picked up a stuffed teddy bear. She squeezed it, making it squeak.

"If I do this [she squeezed the bear], I'm looking for you [she pointed to me] to look here [she pointed to the bear]. I'm showing you a stuffed animal, and I'm squeaking it, and even though you can't hear it, because I want you to [hear]—once you can hear—relate that there's a sound. I know what that is, it's like building a total teaching method. And I just thought that was phenomenal how—I don't know, as a teacher I wonder if I would be able to suspend that 'Well, he can't hear me anyway' kind of thing. But they worked just as hard with him."

The way she understands it is that they have to build patterns of meaning. She told me how she often mimed being able to hear something. She mimicked for me what she saw the therapist do and what she understood she should do at all times. She held out the bear making noise, then took it away. She said, "Then Marianne would be like 'I hear it,' and [Morgan] couldn't hear it, but then she'd bring it back." She was teaching him how to hear by providing "the visual that will eventually go with the oral."[5]

Julia acted out other ways she had been taught to work with her son as part of the therapy to train him for the auditory signal the CI would give him: "Everything was an opportunity—like I hear a knock at the door. Knock, knock, knock! Do you hear it? [She cupped her hand behind her ear.] It's all animated, you know? Getting them excited. I hear the phone ringing! Did you hear the phone ring? [Again, her hand went behind her ear.] Let's go see! [She waved her arm.] Someone called on the phone! If someone called on the intercom, it was like I hear that! Do you hear that? It was all really dramatic."

By engaging in the prescribed EI activities on a daily basis, Julia felt that she was making neural connections in Morgan's brain between an abstract, as-yet-unknown-to-him stimulus (the sound of the door, the squeaking of the bear) and responding and attending to it. She anticipated the CI signal; she laid neural groundwork that would later be paired with auditory information once he was implanted. She was creating neural possibilities by presenting a ghost stimulus and laying the tracks for future neural pathways. This is the daily, sustained therapeutic labor that all parents are expected to perform and is part of evaluating a CI candidate's probability for success. Again, as Nancy told me the very first time I met her, "I don't know that there are many children that the CI did not work for; often it's the case that the implant works fine, but it's the parents who don't do what's needed."

Hearing versus Listening

Once maps are configured after the initial stim date, the CI can deliver more information to the brain. However, hearing this information and understanding it are two very different things. In every single clinical appointment I observed, audiologists used the phrase "good listening." I soon found out that "hearing" is passive, whereas "listening" is the result of purposeful and active *work*. This is the work of auditory training, which results in that good listening.

During a lull in appointments one afternoon, Monica and I sat down in her office to talk. Earlier that day, I had observed an appointment where she had tested a recently implanted child for her response to specific tones and frequencies and asked her mother about her daughter's responses at home to what audiologists call "environmental sounds." These are sounds occurring around us all the time—a doorbell, a telephone, the hiss of a radiator, someone calling your name. "You can hear a whole bunch of things, but you're not going to understand what you are listening to until you start to attach meaning to a sound," Monica said. I offered a scenario of the doorbell ringing. She responded, "Right, and they go like this [she looked around] and [the child] heard it, but it doesn't mean anything. We try to educate parents and say, if you see a behavioral response to a sound, figure out what that sound is, and then show your child what that is. Those are the very beginning tasks to kind of get them, you know, acclimated to listening and identifying."

Hearing is the idea that sound waves travel across space and time and enter the brain through the CI. Hearing does not require *meaning*, because it occurs without conscious neural *work*. "There's a difference between hearing and listening," Monica commented, "because your ear truly is just a receptacle to pick up sound. It's your brain that allows you to understand what your ear is feeding it. It's really your brain that allows you to understand what's being said." This is where auditory training comes into the picture; engaging in this training creates the neural pathways that eventually give meaning to sounds.

Distinguishing between hearing and listening is also how audiologists shift responsibility of "success" from the device to the brain, and thus also to the parents training that brain. Monica explained, "With a lot of parents, we're trying to teach them to start identifying if a child hears something, then start to identify what the sound is, and attach meaning to that." The EI parent training discourse also emphasizes good listening. Jeremy's mother, Carol, had taped EI materials to the refrigerator. There was a paper with instructions for parents to always put their hands by their sides to avoid signing or gesturing. It reads, "Glue your hands to your sides" and "Make your child accountable for listening." As I observed appointments, I found that "good listening" referred not just to the process of attending to sound and attaching meaning to it but doing so without relying on any visual or gestural cues. This second piece is crucial to the politics of sign language, which I expand on in the next section.

The concept of good listening also appears in mapping sessions focused on configuring each of the four programs. For example, when a child positively responds to environmental sounds or tones during the mapping session, the audiologists cheer and say, "Yay! Good listening!" The language of "good listening" also subtly masks an assumption of noncompliance: when a child did not respond, it was routine for audiologists to say, "It doesn't mean she's not hearing, it's just a matter of getting her to listen." Or "Oh, she just isn't going to tell me today." Meaning, if a child does not actively listen, perhaps that is a result of simply being uncooperative. This judgment of uncooperativeness may also be transposed onto the parents.

Finally, the ability to listen indicates the success of the CI. Annette, the chief audiologist, explained that the CI does not work in children who seem to have no concept of sound. One of the first things audiologists test to determine a child's progress is whether the child can recognize environmental sounds or respond to the sound of their name. Some, she said, were implanted but had never been able to do this. "But we put them in the [testing] booth, and we can document that they have access to sound. But it's like their brain doesn't know what to do with this information," she told me, "so we can get what we call functional gain testing and at each specific frequency . . . just tones . . . but they don't seem to be doing anything with that information." In this situation, the problem may be pinned on an auditory processing disorder, which is *not* considered a hearing impairment.[6] In these cases, the question that remains for audiologists is a neurological one. "We don't know if there's auditory processing. We know the information is getting to their brain, but what is their brain is doing with that information—that's the big question mark," Annette said.

Neural Vigilance

All stages of implantation and the notion of "success" are wrapped up in two aspects that cannot be untethered: the brain and language. Two lines of thought regarding good listening and the discourse on neural pathways converge: (1) continuous, consistent auditory training creates auditory pathways for spoken language, and (2) sign language is "risky" because it impedes the formation of these pathways; parents are consequently encouraged to avoid it. Managing neural pathways requires vigilance, as visual stimulus is the easiest sensory input for deaf children. Audiologists and educators build upon the neurological understanding of deafness, reinforcing this message of neural risk. The mantra that sign inhibits speech continues, just in neurological form. For example, Clark, one of the developers of the CI, states, "It is important for a child to develop an auditory-oral language with a cochlear implant first because the neural connectivity demands early exposure" (2003, 695). Furthermore, he added, "this exposure will be impaired if there are opportunities to focus predominantly on visual language and if it is available with total communication" (2003, 696). The basic goal of implantation is integrating the CI's neuroprosthetic functions in a sustained, continuous way, so as to train the child to be able to successfully use spoken language. To achieve this, professionals across institutions (audiologists, teachers, speech therapists) are advised not to let exposure to sign impair these synaptic connections.

One morning I observed a mapping appointment with Kelly and

her son, Nathan, who at the time was approximately three and a half years old. He had been implanted about eighteen months prior, at age two. We all went back to the appointment room, where Monica and Annette hooked up Nathan's CI to the computer sitting on the desk with a variety of different colored wires. Annette sat with Nathan at a table as they played with building blocks; but really he was being trained to move the blocks into the bucket one by one as he heard tones at different frequencies. "We're going to do good listening now," Annette said. While Annette sat with Nathan, Monica sat with her back to them, looking at frequencies on the screen and clicking on different settings to adjust the CI.

As they went through their mapping routine, Monica noted out loud that Nathan was getting services at a deaf school nearby with a CI program. "He sounds great," she said. Monica turned and reiterated to Kelly, "His speech is great!" Kelly answered, "That's what they say." Monica continued to click on the screen. Nathan moved the blocks. I observed Monica and Annette in mapping appointments many times. Because they had worked together for so long, they were able to communicate with each other with just a look. They often conducted these appointments in sync, like clockwork.

Kelly cautiously asked, "So, his school is against signing for him?" Monica nodded her head without turning around. "Yeah, he's oral, I understand why they're saying no. They don't want him to regress." Kelly leaned forward and said, "But when his implant is off, like in the morning before it's on, I'd love to be able to communicate with him, to tell him he's going to school or something." Annette kept playing with Nathan at the table, entertaining him so Kelly and Monica could talk. "You could just do the sign for 'school' with him," Monica said. She then proceeded to show Kelly the sign (although the sign she showed her was wrong—it was actually the sign for "learn"). Kelly attempted to mimic it, and then Monica added quickly (referring to me), "Laura will know. How's his balance?" Kelly said it was definitely better. Monica asked, "Any other concerns?" Kelly answered no, and that she was happy with how well he was doing.

During a subsequent interview in her home, Kelly told me that she had asked about signing with him because in the morning, before putting on his implant, her son asked her questions. She has a two-story house, and when she goes upstairs to wake him, the CI is still charging down in the kitchen. "I want to answer him and tell him he is going to school," she said, but he does not understand her without the CI. "That was the only reason why I thought maybe sign would come in handy," she said. She also explained that signing involves different neural pathways and the oral program he is in at school is against signing because "they feel like he can hear, so they want everything to be just oral . . . so, maybe when he gets a little older and he can . . ." She trailed off, then continued, "I don't want to bombard him too much right now."

Other parents in other appointments raised similar concerns about communication. Usually they noticed their child's frustration at being unable to communicate but cited the strictness of their auditory / oral educational programs as their reason for not "indulging" visual communication. Jane once commented to Monica during an appointment for Lucy, "I'm not ready to see any type of [behavioral] regression go further because somebody decided to draw the line between signing and auditory / oral only." Regardless of how the parents responded about signing, in these situations Monica and Annette typically did not directly address these concerns. They would, as Monica did above, stick to their task of mapping and emphasize the importance of "getting the maps right."

One day I asked Annette, who routinely works with implanted children, about how you know if the implant is *not* working for a child, since in the appointments I observed and the parents I was encouraged to include in my study were those for whom, by all accounts, the CI "worked." She said, "If your criteria for 'it worked' is that they are oral communicators, then [if they aren't speaking] it didn't work . . . and we want them to be oral communicators, obviously." She did, however, reluctantly say that the implant does not always work, and those for whom it does not work should be given access to sign. That is, if the neural processing is not occurring, then there are no linguistic battles to be fought between CIs and sign language. Clinicians often told me that these kids, the "failures," need "some kind of visual information."[7] If auditory neural pathways are simply not going to form, there is no need to disallow exposure to visual language. However, this surrender further reproduces the stratification between the "failures" who need sign language (where failure is located in the patient, specifically the brain, and illustrated through their use of sign) and the successes who do not (where success is illustrated through the patients' performance of spoken language).

"No Sign! No Sign! No Sign!"

The making of neural connections is also highly monitored and controlled because some neural pathways are deemed more acceptable than others. For deaf children, if their neurons were left to their own devices, they would most likely opt for sign language since it is a stimulus that can be presented naturally (unmediated by a neural interface like the CI) and is already socially embedded as language, ready-made for uptake through vision. But often, the use of sign is seen as "the easy way out" or a neuronal crutch; it is also indicative of a parent's *or* child's abilities and/or lack of commitment to auditory training.

Carol, Jeremy's mother, related in an interview that there are "a lot of speech therapists and people out there that really feel that if they don't make it strict auditory/oral only, that the child won't learn [speech]." She lamented the divide in the professional world of implantation. Even though, as Jane mentioned in the first chapter of this book, "the majority of parents want their kids to talk," parents are told that this desire cannot coexist with learning sign as well. Looking back, Carol thinks that it might be "the one thing where we made a mistake when you decide to go this [auditory/oral] route." After implantation, she had enrolled Jeremy in an infant/toddler program at the deaf school tailored for CIs. She had the EI signage up in her house that said, "Glue your hands to your sides! Make your child accountable for listening." She was vigilant, as she had been told to be. "I'm sure you know by now, there's a whole political auditory/oral thing," she said to me. I asked her what this looks like; she described it as "no signing, cover your mouth, no reading lips." Good listening means learning to hear and listen without visual information, including lip reading, she said, "So they said don't sign, don't sign." She asked repeatedly if she should learn sign, and every time the response was "No, don't sign." She had figured that the professionals knew what they were talking about. But now, she said, "I think that was a mistake." Jeremy, implanted at eight months, now suffered at two years from frustration: "He's two, which means he has the desires, the cognitive abilities, the wishes, the motivations, and the goals of a two-year-old. Other two-year-olds say me, mine, mommy, milk, go, no, yes. My baby could say nothing, and he doesn't even know how to form the words, let alone even think about getting them out. But, you know, I was told no, no, no, no! No sign, no sign, no sign!"

Carol was suspicious of the claim that learning to speak must necessarily exclude sign language. She described how she thought long and hard about it, and came to the conclusion that excluding sign was wrong: "I think this is all garbage. This whole political divide between the talkers and the nontalkers. I'm not having any part of it." She did not give explicit neurological reasons for this belief but emphasized the importance of communication for safety: "I'm going to teach him sign for the words I need to teach him. So when he's in the bath, he knows 'sit down.' I can't have him standing in the tub. He knows [she signed] 'careful.' He knows [she signed] 'stop.' "

Carol described how she and some of the other mothers at Jeremy's infant/toddler program at a local deaf school dealt with the therapists and educators. "Everybody spouts the same party line, but some people clearly believed it more than others. . . . Some people are belligerent about it, though. There's a woman, actually, [name], who's a consultant at [child's school]. She's very good, and she really knows what she's talking about. And she's incredibly well respected. But, I have to say, she's not parent friendly at all. The first couple times I met her, I thought, you know, I've got no use for you. I don't need you pointing your finger at me."

She found allies in other mothers who felt the same way she did. They banded together; a few of them together decided that they "weren't crazy about her," and thought of ways to appease her while she was watching. This was how they developed techniques for "dealing with her," Carol said, adding that it was all because the educator would tell them, "You're harming your child" if you signed with him or her. I interrupted Carol to ask her to clarify why she used the word *harmful* to describe exposure to sign. She said, "Because they're not going to be able to communicate [with spoken language] as well as they possibly can and they're going to suffer from that." While Carol understood where this woman was coming from, she said, "I have to live with that child every day."

When I spoke to Becky, Amy's mother, about this experience, she told me something similar. She too described how she struggled to get information across to Amy early in the implantation process. Amy had now been implanted for a number of years and was, by all accounts, doing well in her speech development. However, Becky also turns to sign when Amy is frustrated and unable to communicate. "Once we started learning sign, that was a blessing because then she started to be able to communicate with us," Becky related. She went on to say that Amy did really well with sign and picked it up very easily. "And I loved it, because to me, I was able to communicate with my daughter. I didn't care how I was able to do it, as long as I was able to do it. We finally understood each other. Our frustrations went away. She just was so happy, and I was so happy—and I loved it!"

This is not to say that Amy uses sign language now as her primary mode of communication. She does not. As she underwent the process of implantation, she began to develop speech. But Becky's major concern was that so many parents "were against sign and even against lip reading! They didn't want their child to—they'll do this when they talk [covers her mouth with her hand]—they don't want their child to read lips either. I don't care if Amy reads lips!"

Gray Areas: Rewriting the Script

From the mothers' accounts above, it is clear that the pervasiveness of anti–sign language rhetoric in professional circles does not dictate their behaviors entirely. In the intimacy of their homes, they expressed a lack of total allegiance to one side or the other. For example, after a couple of months of being implanted, Lucy was doing quite well at school in her new auditory/oral placement. In fact, when I went with Jane to a school meeting one afternoon to listen to a Cochlear Americas representative explain how to "get the most out of the CI," Jane gleefully told the other parents and teachers we ran into how well Lucy was producing her "Ling 6."[8] But this does not mean Jane is against sign language. Earlier that day at her house, Jane had told me matter-of-factly where she fit on the language divide. "I'm the parent that says, 'We're OK with gray. We're all right with that.' But there are a lot of speech therapists and people out there that really feel that if they don't make it *strict* auditory/oral only, that the child won't learn quickly. And I kept saying, 'Look, if you're going to learn French as a second language, I don't tell you to stop speaking English just because you're learning French, so you'll learn it faster.' It doesn't make sense to me. So that's where we are right now. We're stuck in this—I want to be gray, and they want to [be] black or white," she said.

Jane's version of plasticity is one that is capable of bilingualism; she believes in the ability to shape Lucy's auditory pathways, and she believes this can occur even in the presence of sign.[9] Jane knows this is not acceptable to the audiologists, therapists, and educators. But she has a strategy. I observed her in audiology appointments with Lucy, telling Monica that she used sign with Lucy and why. She also told me that she decided to be very up front with her child's educational team. She explained to me, "I tell them that I'm doing this [signing at home] and have them work with me on that aspect, as opposed to lying to them. . . . You know what? Yes, we do use sign, that's how we handle our situation." Jane was the only parent I saw directly confront a professional about how to work with Lucy.[10]

Still, the overall culture of implantation is characterized by parents who "don't want their children to learn sign language," Jane said, "Why do you think there's only three kids in that [signing] class? It's the same three that were there last year. [Hearing] parents of deaf kids don't ever learn sign language." She told me she thought this was problematic. Even if parents want to deviate from the strict auditory/oral approach that focuses on good listening, there is no formal infrastructure to do so. Jane told me that Lucy's school is supposed to offer weekly sign language classes for parents, "But nobody takes the sign language classes. So the sign language classes got cut down to . . . I think I had three classes at the school [in a year]." She said she went to the principal and criticized the school for not following through. The result was that she got extra hours of home-based services: "Apparently they'll be giving me some extra hours this year [for sign language instruction]. Five hours a year, so that's five sessions. It doesn't seem like a lot, does it?"

Despite these stories of parents not necessarily following professional advice to consistently and only use good listening to develop spoken language, there are many parents who do. Nancy is a prime example, and the way she frames the experience is one of sacrificing clear communication in the immediate moment for better spoken communication in the future. Nancy strictly adheres to auditory training recommendations, even when it is difficult, she said:

> I didn't doubt [the success of the CI], but my husband doubted it. We were in a store, I remember this, and Anne wanted something. We were going through the aisles in the store, and she just kept pointing and pointing. I said, Anne, that's it, we're going, we're getting out of here. She was screaming. My husband said, 'See? She needs some kind of sign language, she can't communicate!' I said no. I said, 'It has absolutely nothing to do with her communication. She's just stubborn.'

There is tension here, in all of these mothers' stories: the tension between the commitment to neural discipline and auditory training that is asked of mothers for the sake of the child's future, and the immediate communication frustrations that can be solved with sign language.

Precarious Plasticity

In pediatric implantation, deafness is predominantly articulated in neurological terms, mediated by a specific understanding of how the brain works, and framed in relation to the CI, a neuroprosthetic device. Looking at the data in a way that reflects Connolly's (2002) emphasis on micro-interactions—or what he frames as body-brain-culture relays—the neuroscientific becomes more than a discursive feature. Through specific, sanctioned relational techniques of self, CIs are central to and the instrument by which new layers of meaning around disability, perception, and care work are being constructed through a particular form of neuro self-governance.

But I observed a curious phenomenon of contradictory neuroplasticity. On the one hand, neuroplasticity is the foundation of implantation and auditory training. The very principle that synaptic connections can be built with the implantation of a prosthesis combined with carefully crafted exercises is exactly what gives parents faith in being able to create auditory pathways in their child's brain despite the fact that she or he is deaf. But on the other hand, it is a precarious plasticity, even a vulnerable one. The auditory neural pathways are subject to assault, open to attack. The "wrong" stimulus (visual language) could derail them, and thus parents are told not to use sign for fear of causing their child future (neural) harm. This is in direct contradiction with neurolinguistic research for a number of reasons. First, in her review of the field, Baker (2011) outlines the evidence from a variety of studies (e.g., Mayberry and Lock 2003; Mayberry and Eichen 1991) that both sign and speech facilitate mastery of one another. Second, delayed access to linguistic input (such as the time between diagnosis of a hearing loss and receiving and learning to use the CI) and language acquisition of any kind can have negative consequences across a variety of developmental spheres for deaf children (e.g., Meadow-Orlans et al. 2004; Humphries et al. 2012). Third, linguists have found bilingualism—the norm in most of the world outside of the United States—to be beneficial; this appears to hold true in the case of bimodal (visual and auditory) bilingualism in CI recipients (Davidson, Lillo-Martin, and Pichler 2014). But much of the literature written in the fields associated with implantation, such as otolaryngology and audiology, makes quite the opposite argument about bimodal bilingualism in children with CIs. For example, one review in the field of implantation notes that the "primary goal of cochlear implantation is open-set auditory-only speech understanding in everyday listening environments" (Peterson, Pisoni, and Miyamoto 2010, 238). Authors of the review cite plasticity as an area for future research, argue for a neurobiological basis for predicting CI success,[11] and contend that "communication mode post-implantation has also been frequently reported to be a factor that contributes to final speech and language outcome, with oral-only communication producing speech and language results superior to those observed in children who use a combination of signing and spoken language" (2010, 241). And yet a recent study in Iran compared speech development in two groups: children with CIs who have hearing parents, and children with CIs who have Deaf parents who used sign language with them in addition to implanting them and exposing them to spoken language. The results showed that "second-generation deaf children exceed deaf children of hearing parents in terms of CI performance. . . . Encouraging deaf children to communicate in sign language at a very early age, before CI, improves their ability to learn spoken language after CI" (Hassanzadeh 2012, 993).

Two things are worth noting here related to the cultural aspects of science and medicine. The first is that research being done by those sympathetic to the Deaf critique and interested in researching sign language is often seen as "ideological" or outside the mainstream of implantation research. It is also sometimes published only in deaf-specific journals rather than mainstream science journals. The second is that those sympathetic to the Deaf critique and doing research promoting the use of sign language in children with CIs do *not* advocate for sign-only; rather they advocate for bilingualism, whereas those in the fields of implantation often maintain a strict, monolingual, anti–sign language position.

I contend that the neuroscience deployed in the profession of implantation and the discourse about it that trickles down to parents reflect normative values about language and deafness in the scientific fields associated with implantation, not the actual functional limitations of the brain. Whether all deaf children should learn sign language is not the question I seek to answer here; rather my aim is to show how the concept of neuroplasticity is being used to justify prosthesis and promote neuro self-governance, yet it is simultaneously rejected in order to exclude the coexistence of sign language. This contradictory version of neuroplasticity points to the cultural work it does; neuroscience and audiological practice are not acultural, but they are shrouded in medical discourse and thus seem so. Whereas arguments against implantation invoke culture explicitly, the Deaf culturalist critique of CIs emphasizes the role of sign language and culture, as well as the importance of Deaf identity and community. In my fieldwork, I found that many professionals diminish the "culture" aspect of Deaf arguments against CIs and view this critique as dismissible on the grounds of being "merely ideological." Some even refer to the claims of those opposed to CIs as efforts to "indoctrinate" children into Deaf culture. As Nancy told me, "This Audism group [Audism Free America] is older [Deaf] people who are trying to indoctrinate young people." The Deaf discontent here is reframed as purely "cultural" (as opposed to scientific) and, given her use of the word *indoctrination,* even insidious and harmful.

And yet these professionals (and indeed even other parents) deploy certain neuroscientific knowledge to accomplish their own socialization and enculturation processes. The neural arguments in implantation and follow-up care are thus *equally cultural*. But the cultural and ideological work that is accomplished through neurological discourse is done so in the guise of being objective. Exposing the contradictory notions of neuroplasticity as cultural work in the form of scientific discourse stands to be a crucial point on which to build more effective critiques. But it can also be an opportunity for facilitating more fruitful conversations across these contentious divides.

5

SOUND IN SCHOOL

Linking the School and the Clinic

The CI is a technoscientific object with a social infrastructure that stretches across multiple, coordinating institutions that systematize its usage. In this chapter, I show how the social structure and therapeutic culture around CIs that I described in the preceding chapters also appear in schools serving students with CIs. As a result of coordinated efforts surrounding implantation, deaf education increasingly relies on new service sectors, professions, and industries emerging from the CI and related technologies.

I use observations from one deaf education program as a case study to illustrate the cooperation between industries related to new professionalization techniques for educators who work with deaf children, schools, and CI centers. While children receive implants at various ages, educational placement and practices are categorized here as part of the final, ongoing stage of implantation: long-term follow-up care.

The observations presented here are not intended to resolve any of the larger, ongoing debates about deaf education that have been researched and argued over for decades. Instead, my data suggest two significant themes. First, the medical script of deafness is not only adhered to in CI-related education contexts but also is generative of new CI-related industries and professional practices that are used in schools. Second, the neuroscientific understanding of deafness accompanying implantation actually maps onto *and* reproduces the same historic divides in deaf education regarding which language to use as a method of instruction. Namely, as a result of implant technology, there is a new set of stakeholders in deaf education, and the deaf education classroom has been further transformed into an arm of the clinic. Finally, the cultural values surrounding implantation are expressed in an oral education philosophy grounded in the school's ability to produce hearing and speaking individuals who will be "independent," good workers in the future.

Historical Context

Deaf residential schools have traditionally been "manual," meaning that they use ASL as the primary mode of communication both socially and in instruction. Indeed, the history of ASL began when the U.S. educational system became concerned with teaching deaf students. Thomas Gallaudet and Laurent Clerc, two Parisian educators of the deaf, assisted in establishing deaf education and standardizing a signed language of the deaf in the United States. They founded the first deaf school—the American School for the Deaf—in Hartford, Connecticut, in 1817. At that time, sign language was considered to be the "natural language" of deaf persons; the residential school was the primary site for cultivating a common language and the Deaf community (Van Cleve 1989; Burch 2002).

Through the mid-nineteenth century in the United States, ASL was the standard medium for educating deaf students and was romanticized as natural and beautiful (Baynton 1998; Lane 1989). However, in the late nineteenth century, the pendulum swung. Alexander Graham Bell began pioneering alternative methods of education that focused on teaching the deaf to hear and speak; Science and Technology Studies scholar Mara Mills demonstrated that these methods became available largely because of technological changes that allowed for more sophisticated audiometry, or the ability to measure and quantify hearing (2012). After this, "oralism" took hold in American deaf education. Oralism is an instructional method of teaching deaf children that excludes sign language, based on the philosophy that "all or most deaf children should be taught this way *exclusively*" (Baynton 1998, 13, original emphasis).[1] Much of the history of deaf education has been repeated cycles of oralism, then a cycle back toward manualism, and so on. Today, manualism is typically called Total Communication (TC), oralism is typically referred to as "auditory-oral" or "auditory-verbal" (AVT), and these programs are generally mutually exclusive. There are other terms too; deaf educational programs that use sign may be categorized as "bilingual/bicultural" programs, and auditory-oral or auditory-

School data from the Gallaudet Research Institute Annual Survey

	2000–2001 (%)	2009–10 (%)
Deaf students' use of cochlear implants		
Students with cochlear implants	7.4	15
Students with bilateral implants	NA	23.6
Communication mode in teaching		
Spoken	45.4	67.3
Sign	53.2	28.5
Sign interpreter	22.4	14.2
Communication mode at home		
Spoken	71.7	71.6
Sign	28.3	23

Note: The data were published in 2003 (N = 43,416) and 2011 (N = 23,731). Data were not gathered for students with bilateral implants in 2000–2001.

verbal programs are often collectively referred to as "listening and spoken language" (LSL).[2]

One of the most significant consequences of implantation for deaf education is the declining number of deaf children enrolling in TC programs and the increasing number enrolling in AVT programs. Prior to the 1990s—a time when childhood deafness may have been identified at a later age and CIs were rarely used—parents often, though not always, sent their child to residential schools, where the child learned ASL. Parents might also have sent their child to a local specialized deaf education classroom or program if one were available. But "CIs and neonatal screening have catapulted the deaf child into the auditory-verbal camp" (Luterman 2004, 18). In their analysis of demographic trends in deaf education, Mitchell and Karchmer summarize: "It is now becoming much more common for young students to receive cochlear implants. . . . we estimate that, for students 6–11 years of age with severe to profound hearing loss, the prevalence of students with cochlear implants has increased from less than fifteen percent in 1999–2000 to more than twenty-two percent in 2002–2003 (just three years' time!)" (2006, 100). They are including data for non-school-aged children in their 22 percent estimate, but as table 1 shows, implantation in school-aged children increased from almost 7.5 percent in 2000 to 15 percent by 2009. Furthermore, since implantation has become a more routine practice, "some states have documented that parents are choosing the listening and spoken language outcome [educational placement] [in] as high as nine out of ten cases" (Murphy 2009, 22). Compare this to 1997, when only 16 percent of elementary and secondary students with hearing loss were aiming for spoken language acquisition (Murphy 2009). "Today seventy-three percent of elementary and sixty-eight percent of secondary students are learning through spoken language. That's a dramatic shift to occur over just one generation" (Murphy 2009, 22). Many in the Deaf community lament these shifts, and their outcry, covered, for example, in the *New York Times,* has prompted such questions as "Will sign language and the nation's separate schools for the deaf be abandoned as more of the deaf turn to communicating, with help from fast-evolving technology, through amplified sounds and speech?" (Davey 2011).

These shifts are occurring not simply because CI technology is available but rather because of the complex social organization that surrounds this technology. That is, in addition to the therapeutic culture that surrounds CIs, structural and economic factors, such as the development of related industries and highly specialized educational strategies, have emerged from the CI market. For example, new educational and speech therapy methods are offered (for a fee) by organizations like the Alexander Graham Bell Association (AGB), which certifies teachers in new types of oral educational methods that are informed by clinical practices in implantation and a commitment to spoken language acquisition only. The therapeutic culture surrounding CIs—and especially the neuropolitical aspects of it—maps onto long-standing educational divides and serves to rearticulate past arguments for oral education in more sophisticated technoscientific terms.

Linking School and Clinic

Nancy's insights helped me make the link between NYG, AGB, and a local deaf education program. Her daughter, Anne, had been implanted more than fifteen years ago, which was partly what had earned Nancy the moniker of "old-timer." In one of our interviews, Nancy told me how different her experience of Anne's implantation had been compared to contemporary parents' experiences of their children's implantations. She said that when she got Anne her CI, it was the 1990s, and CIs were not as routine then: "Oh, it's not like it is today!" For six months, Anne had to wear an FM system; her hearing aid was linked to a microphone and a transmitter. "There was a whole protocol you had to follow," Nancy said. "Then, you had to exhaust all possibilities. Parents had to go through psychological testing because they said on the street somebody might come up to you and say you maimed your child, what would you say to that? The Deaf community at the time was dead set against it." In fact, she said, if a Deaf person saw you with your implanted child, "They would actually accost you, and say you maimed your child, how could you do that?"

Nancy had status in the CI community because of her unwavering commitment to auditory training, as well as her long-standing relationships with professionals at various CI clinics, schools, and organizations. She had energy; she advocated for children with CIs in a variety of ways and effectively lobbied for support services for deaf children in oral educational programs. She was also something of a guru to other parents. I first met Nancy at the support group at NYG, where she told me that the CI always worked, but the problem was that parents did not always "do what was needed." Another time, I met up with her at an AGB-sponsored parent support group at her local school, which has a deaf education program. She had organized the event, and from the start of the evening, she held court. She was knowledgeable, passionate, and generous with her time and her resources. She provided guidance and advice to the "newbies," parents with newly diagnosed or implanted children, once again illustrating that attending to parents' emotional needs breeds community and institutional connections. On any given day, Nancy is strategizing ways to make implantation better, to make the community better organized, and to make educational programs and their links to clinics stronger.

I arranged with Nancy for her to escort me to a CI-tailored Auditory Verbal Therapy (AVT) program near her home. I chose this site because Nancy was a natural point of access to it while I was in the field, and also because it had recently received some press coverage. This coverage describes it as one of the nation's strongest programs in oral education of deaf students.

Nancy is a white woman in her early fifties; her home is nestled in a middle-class, residential, and suburban area near New York City. When I arrived at her home, the phone started ringing as soon as she ushered me in the front door. It was one of her colleagues at AGB calling. They were collaborating on a CI-related conference that parents and professionals could attend. She discussed the information for the brochure and hung up; the phone promptly rang again. She discussed more details with another colleague, then told me about her AGB community listserv and regional contacts in various undergraduate programs for audiologists and speech-language pathologists. She sometimes visits these programs to give presentations on how important oral education is, emphasizing that not all deaf children need sign language, as many seem to think. I asked her how she knew so many people and had so many contacts. "When you've been in it this long, you know everyone," she said.

I asked her to explain further how interconnections between educational programs and the clinic happen. She commented, "It depends. Sometimes Sonya [the NYG social worker] will call me and say, 'Can you speak to this person?' A lot of them meet me at the support group. From the school, Linda [the principal] will tell parents to call me." She explained how important it is that the school double as a site for AGB-sponsored dinners (which I described in chapter 2) for parents: "We have AGB meetings at the school because most hearing impaired kids will get services from these schools," she said. "I'm at the dinners that we do, I'm at the meetings, I'm at the open houses, you know, so . . . and even some of the teachers here have called saying they have a family that has some questions, 'Would you mind speaking to them?'" Nancy also works as a fundraiser for other schools that provide oral education methods. We continued our conversation as we drove to meet Linda, the principal (who was also active in AGB). As we pulled up, Nancy said, "It's a pretty small community."

The school looked like any other elementary school, with colored pieces of paper on the walls with drawings, and teachers' voices spilling out into the hallway as they conducted their classes. There are eighty children at this elementary school, and more than half of them have a CI. The Department of Education is trying to prepare for the continually rising numbers of students whose parents want these kinds of services, and this is one of the programs they have been paying attention to as they develop ways to cope with the demand. Nancy led me to Linda's door, and the two of them greeted each other warmly; they hugged and said hello with an easy rapport. Nancy introduced me to Linda. The three of us walked through the hall, peering into rooms, as they both explained the setup to me.

One of the first things Linda told me is that there is an increasing number of children with a CI on one ear and a hearing aid on the other, and also that the incidence of bilateral implantation in her student population is rising. This increase in her student population mirrors a broader trend toward bilateral implantation, which has been confirmed by researchers (e.g., Brown and Balkany 2007). Indeed, in this school, students increasingly use hearing aids and CIs together. The school audiologist said that they have more kids with CIs than hearing aids and that this increase has occurred in recent years.

All of the children in this program had hearing aids and/or CIs and were using spoken language. In the first classroom we came to, we stopped and observed the teacher, who had a variety of empty plastic film canisters filled with different items. The task at hand was to shake the canister and based on the sound it made, determine what was inside. There were about eight students in this room with CIs. I watched as one implanted child—through spoken language—guessed correctly: pennies. I immediately noticed that the teacher wore a microphone, and the room had speakers mounted at various points, approximately one on each wall close to the ceiling. In the 1990s, oral programs depended solely on sound field systems, which consisted of a microphone, a "base station," and multiple speakers. The teacher wore a microphone, which took the acoustic signal and sent it to the base station, which then transmitted the signal to the speakers located throughout the room. This ensured that the auditory information was distributed equally to all areas of the room, thereby immersing all the children in the "sound field."

Linda explained, "About five years ago, they switched over to personal FMs. . . . Now a personal FM is one where the teacher is wearing the mic and the children have their own FMs in their implant." These personal FMs, which operate through radio waves, can be used with either hearing aids or CIs. For the children in this classroom who had CIs, an individualized signal was transmitted directly into the child's CI microprocessor through a special jack that is mounted on the CI's BTE unit. Linda said, "So you're seeing both—you're hearing through the sound field, but the children also have direct input [to the auditory nerve]." This individualization of FM technology through the use of wireless jacks mounted on the CIs means that the classroom equipment can be tailored to meet each student's particular audiological needs simultaneously. That is, the maps in each child's CI, which are defined by medical professionals, take the signal from the sound field and translate it into a set of ones and zeros that reflect the features of the child's own specific hearing loss. "So it's very tailored . . . that's going to depend on the type of hearing loss they have and the audiological recommendations," Linda continued, "I was happy we were individualizing, and that we had the sound field, and each child had his own system."

CI Center and School Partnerships

Because of this individualization, each child's specific hearing loss has to be medically monitored in the classroom. This is why local CI centers play such an integral role in this CI program; in order to medically monitor and customize all the CI-related assistive technologies, CIs need to be mapped to the educational space. Recommendations for classroom adaptations and CI configurations for each student come directly from the clinic, turning the school into yet another arm of the clinic and extension of medical practice. Hence, the classroom is a *prescribed* environment. Some implant centers will "send their people here to do mappings for the kids," Linda said. This collaboration is one of the program's pioneering efforts that has caught the attention of so many in the field of oral education for the deaf. "So you can choose to—if you were implanted at NYG, you can choose to go there to have it done. Or you can make the arrangements when the person comes here, to do that," she described. Audiological services at the school are sophisticated: "These are the actual audiologists from the implant centers. They schedule time to come here, and they do the mappings onsite. . . . It's a partnership."

When we walked into the audiology office at the school, the first thing I noticed was the entire wall full of tiny clear plastic drawers. These drawers were filled with a vast number of variations of cables, coils, jacks, spare parts, electro-cords, electromagnets, and batteries. Linda explained that the audiologists have to be able to replace any part of any child's hearing aid or CI at any given moment. She reflected, "It used to be, everybody had one [hearing aid] ear, and you just made the molds and that was education of the deaf. . . . That was the old style of deaf ed." But when Linda took over this program, the first thing she instituted was a full-time audiologist so that every child could have his or her personal system, which "is a prescription from the center audiologist." She wants the school to be much more precise and much more tailored to the students, so that their educational/communication method reflects their medical prescription: "Obviously it meant a huge paradigm shift, from a generic application system to customizing the FM or sound field or systems to the hearing aids or the CIs . . . and the whole idea is that you want it to be customized as precisely as possible."

Staff crossover is not the only cooperation between implant centers and schools. Economic links between the clinic, home, and school are sustained in other ways. For example, CI companies like Cochlear (as well as Med-El and Advanced Bionics) fund organizations like AGB in various ways, such as by supporting their annual conferences, among other initiatives and programs. Archbold (2006) outlined the many organizations that work together as liaisons between CI corporations, education professionals and associations, as well as health care entities in the education of children with CIs. Thus, my observations of Nancy—who volunteers her time with AGB and participates in many activities, like providing parents with community through dinners, information sessions, and smaller regional conferences, as well as organizing parent volunteers to reach out to other parents—are not anomalous but rather a product of larger economic cooperations.

The CI team meetings I sat in on also reflected these cooperations, where professionals termed "educational consultants in a medical model" worked with NYG on each child's case. These consultants visit the children's current educational placements to assess the accommodations. This assessment is part of the rubric used in determining CI candidacy and ongoing efficacy of the implant. In these CI team meetings, each child's school placement is discussed in depth.

Teachers of the Deaf

As children pass from one grade to the next within the CI program, they are eventually integrated into the "regular classrooms" that share a building with the CI program. Linda labeled these "collaborative classrooms." The idea behind these classrooms, she explained, is that there are two teachers assigned to the room, where a mix of hearing and implanted students are taught cooperatively; that is, Linda clarified, it is "a mixed classroom with both a teacher of the deaf and a regular education teacher."

The term *teacher of the deaf* is an entry point to illustrate how implantation has not only generated new stakeholders and educational industries but is also expanding and/or redefining professional roles. Historically, teachers of the deaf are teachers trained in ASL so they can work with deaf students. Thus, when I first heard the term *teacher of the deaf,* I imagined one who would be easily identifiable by his or her use of ASL. When we observed the first classroom, I saw that there were indeed two teachers in the room. However, no one in the room was using sign language. All of the children appeared to use spoken language to communicate at all times in the classroom, as did both of the teachers. While observing the collaborative classroom in action, I was unable to distinguish between the children with CIs and the hearing children. Additionally, I could barely distinguish between the teachers of the deaf and the regular education teachers.

As we stood in the back of the room, Nancy and Linda beamed at me, watching me as I tried to figure out which teacher was the teacher of the deaf and which one was the regular education teacher. Linda explained to me, "See, they [the teachers] are sharing and supplementing and modifying and accommodating and doing all the things to make this a successful placement [for a student with a CI]." Linda noted, "Many times, when I'll point these classrooms out, I'll ask, 'Who do you think is the teacher of the deaf?' . . . And 99 percent of the time they don't know." In a collaborative classroom, the CI provides a student access to spoken language, with the educational infrastructure supporting it that can enrich and reinforce this language constantly. To illustrate, Nancy explained:

> One time I brought someone here, when [her daughter] was in second grade. The teacher of the deaf was explaining something . . . and as she explained it, she said, "OK. Now we're going to look at this door. And I'll just give an example. The door is open. What's another word for 'open'? 'Ajar.' " . . . They just pull out much more vocabulary—they make the connection with language, so that in reading, in math, whatever they do—it's really language enriched. . . . The teacher of the deaf also makes sure that when a certain concept is being taught—maybe she'll have visual supports, like charts and pictures and things . . . and add more to the multisensory feel of the classroom.

In this context, students are immersed in what Linda and Nancy refer to as an "integrated environment," and they benefit from a smaller student-teacher ratio. Not only are there two teachers in each classroom, but "the curriculum is the standard state curriculum, with just whatever legal mandates are available for the hearing-impaired and deaf, within the classroom—extra time on tests," Linda said. Nancy spoke up here with a point that she stressed as extremely important to her: "They do not dumb down the curriculum for these kids. My daughter takes the regular, standard state test. . . . They're not saying, 'Oh, poor deaf kid.' You know?" She continued, "In the old days, [deaf students] were so segregated, and there are still programs—like [local deaf school that uses sign language]. When you go into a classroom there, the teachers of the deaf have no idea what the general ed expectations are for the children at that grade level because they're not immersed in it. It's not their fault."

The teachers of the deaf who I observed in the collaborative classrooms are teachers who have been trained in specialized deaf educational practices known as listening and spoken language (LSL).

LSL certification can be obtained from the AGB Academy for Listening and Spoken Language, which is a corporate subsidiary of AGB, established in 2005. Their goal is to train enough deaf educators, speech language pathologists, and other support service providers in LSL techniques so that a certified professional is available to students in all geographic areas across the United States. According to their website, "The AG Bell Academy for Listening and Spoken Language is the global leader in certification of listening and spoken language professionals" (http://www.listeningandspokenlanguage.org).

There are two types of certifications available to become an LSL Specialist (LSLS). The first type is the LSLS Certified Auditory-Verbal Therapist (AVT), which certifies those who work in one-on-one therapy sessions with children and their families. The second is an LSLS Certified Auditory-Verbal Educator (AVEd), which is for those who work with children in individual, group, or classroom settings. Thus, the speech therapists who work with students outside of the classroom are typically certified in AVT, while the teachers of the deaf I observed in the collaborative classrooms are certified as AVEds. AGB partners with and receives funding from CI corporations. They certify various professionals who work with children with CIs and who collaborate with CI companies, clinics, and schools. These certification programs for the professionalization and specialization of teachers to work with children with CIs are just one example of some of the new markets that implantation practices have created.

The Linguistic Divide: "Completely Separate"

Neurological arguments over "real estate" in the brain directly correspond with the school buildings' physical layout and the school's conceptualization of the relationship between the oral and manual education programs for deaf students. That is, at this school, the TC and auditory/oral programs are professionally, geographically, and ideologically separate. The building containing the CI classrooms and collaborative classrooms is literally separated from the TC building by a street and a five-minute walk across a field. As Nancy, Linda, and I walked over to the TC section, Linda explained that the TC section did not employ the same technologies or methods that we had just seen in the CI and collaborative classrooms. She emphasized to me that the programs did not have crossover: "They are completely separate."

There were about half as many students in the TC program than in the AVT/CI program. Linda recalled her experience of setting up the programs: "I was doing in-service with staff from auditory/oral, and the TC staff felt like they were left out. I would provide in-service for the TC staff, and the auditory/oral teachers would feel left out. So I recognized that you really cannot have both modalities in one program."

But the separation of the programs is an administrative decision that reflects assumptions about the neurological capacities of the implanted children versus the nonimplanted children. For example, the way this school program is organized reflects larger professional assumptions that children with CIs in the collaborative classrooms would be harmed by exposure to ASL. Interestingly, even though pediatric implantation in the United States continues to increase, the number of students in the TC program at this school also continues to increase. When I asked Linda to explain this, her answer partially reflected that her program was in proximity to New York City. She said, "We have so many families coming in from other countries. . . . And the children are seven and eight years old. . . . They have missed the auditory development stage, which is zero to six. That's the critical stage for learning to listen through the auditory channel. If you miss any portion of that, it's very difficult to catch up . . . it's very difficult to expect a child of seven and eight to catch up auditorily. So, we give them sign language."

This echoes two points from chapter 3. Firstly, as Annette demonstrated in the chapter's discussion of candidacy, professionals in implantation acquiesce that some children are or will be CI failures because of biological factors (e.g., age) *and* social status or cultural background. As a result, they believe that there is no reason to deny exposure to ASL for children who emigrate at age seven or eight; their brains are already a "lost cause" since their age has impeded the process of auditory and neurological training. Furthermore, children in immigrant families may not be seen as ideal CI candidates in the first place due to the language barrier. Professionals see the ideal CI candidate as one who is as young as possible and parented by English speakers.

The social organization of deaf education is a direct outgrowth of clinical thinking, particularly in relation to the neurological narrative I have described in the previous chapters. Finally, while the separation (and even hostility) between TC and AVT programs is not new, what is new is how it is being rationalized systematically in relation to neurological capacities and CI technology.

Supply, Demand, and the "Real World"

In an interview with Nancy shortly after visiting the school together, we talked about the controversy and opposing views over which language to use when educating implanted children. She brought up the topic by telling me a story. She had recently attended a panel discussion with various professionals and educators on the topic of CIs and sign language. "Why would you get an implant and still sign?" she asked. She was baffled, "I could see all the professionals down in the bottom row who were still supporting sign." She explained that she had no idea what use sign would be if a child had a CI: "They said, sign is the natural language of the deaf." She then told me that Linda, the principal of the school, her friend, and a CI user herself, had attended this event with her. When advocates for sign said that sign was the "natural" language of the deaf, Nancy said that Linda looked over at her and angrily told her that English was her natural language, not sign. "Cultures evolve, and with technology it evolves faster," Nancy added. "So yes, you have a culture, but why not integrate everything into that culture if it's something that can make the culture better? It's not making the culture worse, it's making the culture better. . . . Don't they understand their numbers are going down?" Nancy asked.

Furthermore, she stated, more and more parents are demanding that schools be able to provide CI-tailored education, such as through the program we visited, and LSL-certified teachers. "This is a case of supply and demand. Parents are demanding this, and you're going to have to supply it," Nancy said. Earlier that day, Linda had also explained to me that "in life, as an adult, when it comes time to get employed, that is going to be their gift. The fact that their speech and language and their ability to converse is one of their strongest areas—that could have been taken away, if they had not had that option [to learn spoken language]." Nancy also shares this perspective: "My child is going to learn to deal with what we call the real world—because you can't have an interpreter next to you all the time and you can't live in that little community all the time. You have to get out."

As children move through the postimplantation stage of long-term follow-up care, educators perpetuate through oral education programs the same therapeutic culture instilled in families throughout the prior stages. One way this therapeutic culture translates into educational settings is through fully embracing the values of the hearing world. "The way that the CI school program is run, the philosophy, what they do with the kids, how they move them along, no hand-holding, no coddling. This is the real world, " as Nancy had put it. And I had heard this sentiment about the "real world" echoed throughout parent interviews as well. Carol, like all the other parents in my study, spoke of this "real world" element but usually talked about it in relation to opportunities. "The language of our world right now is spoken. I want him to have opportunities, I want him to have that spoken language," Carol told me. Carol and Jane both talked to me about the increasing enrollments of deaf children in AVT programs. "It is going up, up, up, but the enrollment for the TC kindergarten through eighth grade is going down, down, down," Carol said.

Carol and Jane expressed sympathy toward the Deaf community regarding these changes. However, they, and other parents, repeatedly stressed the "opportunities" they sought to provide their child by having them learn to speak. Every imagination of their child's future was in terms of his or her ability to hear: their ability to be educated, successful, independent, and happy. Because, to them, this all depends on the ability to hear, the child's future hinges on implantation and the task of successfully acquiring spoken language. Morgan's father, Paul, expressed a similar sentiment. When talking about why he felt strongly that getting a CI for Morgan was the right thing to do, he said, "Let me ask you a question: Suppose he's out with his friends and he tries to hail a cab in the city. Could he do it? No. Could he tell the guy Forty-Fourth and Broadway? No. Could he hear what the guy is saying back and forth? No, he can't. He would need to hear in this world, just like you do, you can hear in both ears. Get a CI and you have hearing."

Nancy used market discourse to talk about how families navigate institutions and advocate for CI programs for deaf children. At the heart of the "demand" side of the equation is the desire parents have for their child to succeed in the "real world." This conceptualization of the "real world" and wanting to provide their children with all the opportunities possible not only reflects middle-class parenting styles and concerted cultivation but also an assumption that d/Deaf people who use sign language to communicate experience ineptitude and isolation. This conceptualization incorporates considerable imagination about deaf people being unable to communicate with people in everyday contexts (a taxi driver, for example), which is generally easily worked around: Someone could easily type an address on a phone screen and show it to the driver or perform other such minor adaptations. This conceptualization also sees sign language as a limitation or social restriction rather than a shared language and entry point into other communities. Finally, this conceptualization destabilizes the idea of what is a "natural" language for deaf children in communities where hearing and listening are valued, because technologies like the CI mean that what is "natural" for deaf children is actually flexible and carefully coconstructed through social interactions *and* technological artifacts (Law and Hassard 1999).

The general ethos toward signing that I observed in the CI community is that sign language renders deaf people unable to be "independent" in the world and that spoken language is liberatory. There are indeed far fewer people in the world who know sign language than spoken language, and concerns about the ability to communicate with anyone at any given time are legitimate. But the assumption that deaf persons cannot function in day-to-day life goes beyond such a concern and reflects broader ideology about disability and difference. As sociolinguists studying language promotion, languages of marginalized groups, and language suppression have noted, the status of a language often indicates the status of the group who uses it (Milroy 2001; Garrett 2010). "It is generally difficult to distinguish attitudes to language varieties from attitudes to the perceived groups and communities members who use them. Language varieties and their forms are often not simply characteristic of a community, but even enshrine what is distinctive in the community and in a sense 'constitute' that community" (Garrett 2010, 16). In other words, assumptions about sign language cannot be separated from assumptions about what it means to be a Deaf person in the world.

Finally, just as sign language is a core cultural value of the Deaf community, being able to listen and use spoken language is a core cultural value that binds the CI community. CIs and spoken language are seen as progress from the unmediated forms of deafness characterized by the use of sign language. Sign is analog, outdated—indeed, indicative of a technological ghetto—and relegated to use by those who are not committed to or are unable to succeed at auditory training. But the meaning of sign and the meaning of spoken language in culture are not a given. Rather, they are socially constructed and based on hierarchies of bodies and abilities.

Deafness as a Thing of the Past?

The educational placement of a child after implantation is ongoing and long-term. The therapeutic culture of the clinic and EI translates into the school setting; families demand CI-related services, new professional sectors and economies emerge, and the resulting community of parents and professionals coalesce around the cultural values of listening and speaking. In sum, the goal of implantation and the accompanying educational programs that service implanted children is not to socialize a Deaf child. Rather, the goal is erasure of the "deaf" qualifier entirely, particularly through training the child's brain in *only spoken* language. That is, the neural project of implantation involves working to mitigate deafness to the point that it becomes no longer able to be perceived by others. As Linda told me that day while observing the implanted children in their collaborative classrooms, "We think that deafness is going to be a thing of the past someday in the future." However, because of disparities in implantation and the immigration of deaf children who are "too old" to be deemed CI candidates, the disappearance of deafness may be feasible—if it turns out to be true—only for white, middle-class children.

The integration of medical devices for hearing into the classroom is not a novel result of CIs. Previous oralist, medicalized educational strategies tried similar technologies (i.e., Mills 2012). However, emergence of the CI has resulted in a new, neuroscientific, and *customized* iteration of classroom integration, characteristic of biomedicalization. Like other types of enhancement technologies, the CI's ability to be customized is based on it being seen as "more powerful, precise, targeted, and successful—powerful because they are grounded in a scientific understanding of bodily mechanisms" (Rose 2006, 20). The discourse of biomedicalization creates faith in the CI's precision and success; this discourse relies not just on simple normalization techniques but also on complex systems meant to achieve customization. But indeed, the CI appears to "work"; as I observed in multiple classrooms at the school, implanted children spoke and interacted like any other hearing children. Although I do not discuss outcomes or the efficacy of these programs here, they are certainly a rich and needed site of research and analysis. But what is clear is that the orientation toward CI technology, the promotion of customization, and the emergence of related markets result in classrooms that blend seamlessly into the ideologies of the clinic.

Conclusion

THE POWER AND LIMITS OF TECHNOLOGY

> *As "choices" become available, they all too rapidly become compulsions to "choose" the socially endorsed alternative.*
>
> • Ruth Hubbard

One afternoon, Jane and I drove to Lucy's school for a parent workshop that was being given by a representative from Cochlear on how to maximize the CI's potential with tips on operating the equipment. Now that Lucy had gotten her CI and was in the auditory-verbal classroom, Jane was taking advantage of every opportunity to learn how to use the device. On the way, she told me about other resources at the school. For example, an ongoing parent group trained parents to work with their implanted children. She called it a "training ground," because the group teaches that "everything is an opportunity to do language development," such as making crafts at home or getting ready for holidays. She continued, "What I've learned from [the group] is, I don't scream at my kids to get the hell away from me because I'm making dinner, but I include them in on what I'm doing to increase the language opportunities that are available. That's the core of everything . . . during bath time, cooking time—bring them in, have them help. It's starting to become very natural. I don't just ask where the toothpicks are. I say, 'Lucy, go over and ask that man where the toothpicks are,' and I'm behind her to correct her. Plus, it's socializing her, stepping out of her shell, learning complete sentence structure."

When we got to the school, we joined a room full of women. About ten mothers and a couple of classroom professionals who work with children with CIs all sat around with notebooks and pens, ready to hear what the Cochlear representative had to say about switching between programs, adjusting settings, maximizing battery life, and the like.

Later that day, after the meeting, we talked about what it was like to raise a child who is different. Jane concluded that when you have a child with a disability, or who is different in some way, the automatic response from other people in the world, she said, is judgment: "If there is something wrong with my kid, then there's something wrong with me." She went on to say that people do and will continue to look at her and judge her. "This is based on years of parent support groups . . . but the impression I get from many parents is that their children are a direct reflection of them and that people will judge them, and people don't want to feel judged," she commented. Norms concerning disability and difference, the availability of biotechnologies to mitigate disability, and the expectations put on mothers to utilize medical interventions are powerfully intertwined.

The parents in my study are committed to auditory training and accept the therapeutic culture of implantation, but that does not mean that they think of their children as hearing. All of the parents have reiterated to me in interviews that they understand that their child is still deaf, even though he or she has been implanted and is using spoken language. Kelly—who had asked Annette and Monica about whether she could sign "school" to her son Nathan—reflected on her understanding of her son's deafness. Nathan had been implanted for more than two years, and his speech, by all the audiologists' and teachers' accounts, was excellent. She said, "[The people at] NYG are right—he'll just grow up to be a child with a hearing loss. It's not going to be who he is. He's just a child with a hearing loss . . . if he asks me, I'll be upfront. You're deaf. That's fine. I'm not trying to undo the fact that he's deaf. It's just that I'm trying to help him. If I can help him here, then why wouldn't I? I'm not ashamed of it. I just did this to help him."

Jane emphasized that Lucy had not been made into a hearing child by being implanted: "OK, yeah, you fixed her hearing. And? What about when she loses the equipment? How about when the batteries are dead, and it's gone for four hours? How are you talking to her? I am still very proud of myself for maintaining the sign language. We are still using it at home. There are times when I absolutely am not, but when I need to clarify something with her, I go back and, yes, I pair [speech and sign]. I never stop talking, ever. Sometimes I'll talk without the sign and I'll only use the key words [in sign]. . . . I have to worry about safety. I still have to raise a child. I understand that she has to learn to speak. But I still need to be able to sign 'stop.'"

For the parents who had experienced all five stages of implantation and were now living through the years of follow-up care and maintenance of the CI, having their child use the device, managing its functionality (such as battery power and reconfiguring the maps as needed), and engaging in constant auditory training had become part of their lives. The families in my study were largely compliant with the expectations of professionals. The technology and its demands were integrated into family life and child rearing, illustrative of what Timmermans and Berg (2003) suggested is a "subtle restructuring" of identities that occurs because of the medical technologies available today. However, the children in my study have mostly been just placeholders—talked about, referenced, and for whom decisions are made—and yet we know little about how they will feel about the CI and their experiences as deaf children. Future studies of children with CIs as they grow older will be crucial to undertake.

The Power—and Limits—of Technology

The experiences of the families in my study are partially shaped by different viewpoints, or scripts, regarding deafness. Some who adhere to the Deaf cultural script believe that being deaf is more akin to belonging to an ethnic group with its own language. They may identity as Deaf and argue that the Deaf community is best understood as a unique minority linguistic group. They categorize deafness as a social difference better understood within the broader framework of diversity. They also believe that deaf children should be communicated with in sign language because it is accessible to them and a ready-made, well-established language with a thriving community and culture. From this viewpoint, children should be signed to because sign provides an immediate access to the world around them, a way to communicate without medical or technological intervention, and a conduit for fostering Deaf identity and community. This view rejects deafness as a problem that requires medical or biotechnological intervention. It holds up sign language as a social technology that answers any challenge that deaf persons might face. It deplores—though not universally—hearing parents' pursuit of pediatric implantation.

By contrast, the other view of deafness is as a condition that warrants medical and biotechnological intervention, and the embedded goal of this intervention is to render the condition inconsequential. Professionals and hearing parents, who represent the majority of those who work with and encounter deaf children, acknowledge the existence of a Deaf community; however, they most likely do not integrate or agree with the community's views on deafness. Instead, the goal is to overcome deafness through a CI, to render it as invisible as possible, and to achieve this through the accompanying therapeutic child-rearing practices. To explain deafness, they turn to scientific discourse and medical knowledge, rather than the Deaf community's discourse of Deaf identity and community. Those who subscribe to this medicalized script of deafness may not accept sign language as a viable option for communication, unless implantation is not feasible or successful. While they acknowledge sign language as a full language, they generally see it as compensatory, neurologically risky, or indicative of a lack of commitment to or success of auditory training.

These different scripts of deafness mark the contours of the political divides over the technology of the CI. Those who adhere to the Deaf cultural script are less likely to support the use of CIs, while those who adhere to the medical script of deafness are more likely to support them. These two sides of the debate over CIs have often played out in polarizing and antagonist rhetoric, although it is certainly not a black-and-white issue; people may take a position somewhere between these two extremes of the debate. In this book, I have focused on the features of parents' experiences and the social technologies used by professionals in pediatric implantation to ensure parents' compliance, rather than proceeded from the starting point of trying to determine whether the CI is a good or bad technology. By stressing the importance of family context, and the pressures on mothers in particular, I have shown how families' experiences of CIs are shaped by the politics around them but are also quite different from and more nuanced than the political debates—and even challenge the strict narratives of compliance that professionals advocate. Professionals may encourage strict divisions between sign and spoken language, and thus between the CI and the Deaf community, but parents sometimes incorporate features of both scripts. Furthermore, the five stages of implantation that occur over numerous years—identification, intervention, candidacy, surgery, and long-term follow-up care—are marked by anxiety, therapeutic labor imperatives, neural narratives, and hope for the future. Implantation is not a onetime decision to adopt a device but rather an ongoing negotiation between the different sets of expectations that accompany raising a deaf child.

Given the distance between the political debates and the patterns of the families' experiences in my study, I do not fully accept the terms of either side of the debate and challenge both viewpoints. Both sides of the debate focus overmuch on the technology of the CI. To begin with the pro-implantation side: Some who support the use of CIs may overstate its capabilities or maintain a tremendous amount of faith in the device.[1] But as I have shown throughout the book, social factors play a greater role in outcomes than technological prowess. As such, these factors deserve more scrutiny, especially given that the outcomes of implantation have been routinely documented as highly variable (particularly related to what Chang et al. [2010] called "downstream disparities"). And to turn to the anti-implantation side: Deaf culturalists have expressed skepticism about the technology since its outset. However, some have claimed that opting for the CI *determines* a child's community membership in the hearing world. But in this book, I have shown that CIs do not determine who a person becomes; rather, social technologies (such as the anticipatory structures and therapeutic culture that surround the CI) and structural position (like race, family status, ethnic background, or class) collectively shape individuals' experiences and identities alongside the technology. These are social and cultural arrangements that have been made, and depending on one's view regarding the power that individuals have to shape society and vice versa, they can be unmade or reconfigured. Deterministic claims about the CI—either as a miracle device that will eliminate deafness completely or an evil technological artifact that will destroy Deaf culture—should be approached with caution.

The Cultural Work of Medicine and the Ideology of Ableism

Professionals involved in implantation actively and purposefully socialize parents into a set of values concerning hearing and speaking. They also sometimes dismiss the Deaf critique as merely ideological, even as they are making ideological claims of their own. As a result, medical advice is consistently reified as objective and acultural precisely because arguments against implantation are cast as merely cultural phenomena that undermine the goals of medicine. But the intervention of this book is to show that the socialization process in implantation and the norms regarding CIs in the medical sphere are equally cultural; it is just that the cultural claims are more in line with dominant social norms concerning disability. It is often hard, for a variety of reasons, to see medicine operating as such. Most people are unaware that scientific claims and medical knowledge are also cultural; for example, I showed that the pro-implantation arguments that make claims about the neurological capabilities of children's brains reflect cultural values, not actual, neurobiological limitations. It is easy to forget that medical and scientific knowledge is cultural; it is shrouded in objectivity, and beliefs about disability circulate in its shadows. And while I make no claim as to the goodness or badness of CI technology, I do reject the claim that medical or neuroscientific knowledge is acultural or ahistorical. Instead, I claim that medical knowledge about and neuroscientific explanations of deafness are as *equally* cultural as the Deaf viewpoint. And the cultural work of medicine and the cultural de/valuing of disability are two sides of the same coin here; normative values regarding hearing and speaking that undergird implantation practices are exactly that, value systems that we take for granted to be "common sense."

However, the Deaf critique too narrowly argues for making decisions about deaf children based on Deaf experience and fails to take into account other, powerful social processes occurring for families with a deaf child. It is not that Deaf experience should *not* be considered an authoritative account of living with deafness. It should. But by *only* relying on Deaf identity politics, the Deaf critique fails to incorporate women's parenting experiences and acknowledge the pressures that society puts on women to engage in scientific motherhood in order to perform as "good" mothers. Deafness occurs within the context of family systems, and there are many ways in which a child experiences deafness and grows into a deaf adult. As it currently is articulated, the Deaf critique does not always adequately tend to the grief that families experience when their child is not what they expected, particularly when that expectation pertains to the experience of disability. On this latter point, the Deaf critique of implantation has been mostly unwilling to engage with the politics of or theories about disability on a broader conceptual level. This is not only ableist but also an impediment to productive dialogues with communities of parents and disability rights allies. In short, I am arguing that both sides of the debate over CIs and both scripts about deafness are affected, albeit differently, by an ideology of ableism.

The ideology of ableism and its will toward normalization enhance the power of medicalization and the seduction of a technological fix. And it is on these systems of meaning that I focus my critique in this book, not actions of individuals or families' choices. That is why throughout this book, I primarily focus on the power that the process of medicalization has in facilitating parents' adoption of the medical script of deafness. To do so, I show specific structural and cultural aspects of this socialization. Established social technologies shape how parents receive and interpret information about their child's deafness; *anticipatory structures* in the clinic inculcate them into a particular way of thinking about deafness; neural narratives about building the correct synaptic connections structure daily interactions with deaf children; ongoing institutional cooperations provide emotional support and ensure compliance; and emergent professional markets in the education sector extend clinical thinking beyond the clinic and into schools.

Ambivalent Medicalization

While this book examines the particular case of CIs, it also addresses larger questions about society's relationship to medicine and medical knowledge. Most studies of medicalization focus on making the case that medicalization is powerful and positions the individual as subject to it. Others may characterize medicalization as a "neutral process." Undoubtedly, the power that medicalization yields over individuals was clearly described in this book. But this relationship to medicalization is not as simple as individuals being overpowered by a larger system. I would not make that claim; it is more complicated than that. In a highly medicalized society, we both surrender to and may be empowered by medicalization. That is, medicalization is not neutral; it is ambivalent.

One of the ways this ambivalence is illustrated is by looking closely at disability within a family context and acknowledging the range of experiences mothers go through. During the stages of implantation, parents experience many emotions, including relief from knowing they are participating in interventions, hope in the technology, and even community through their connections with parents who share their experience. Yes, there is grief in diagnosis, but there is also relief to be had when you are given a plan for action. There is faith in the interventions but also doubt and at times a need to resist, to sometimes use sign language despite being told not to. The therapeutic labor also sometimes pays off, and mothers experience social rewards when living out the narrative of overcoming deafness. And there is no doubt that many mothers were proud to be a part of the CI community and even, as Jane and Nancy did, experienced deep self-fulfillment through mentoring other parents. Meanwhile, national organizations and support groups provided spaces for children with CIs and their parents, generating what some might call a "CI culture." While I showed mothers' willingness to perform the labor, I also showed the imperative to do so that the clinic places on them. While mothers were socialized into one community, they were also told to reject another. And who is to say one community is "better" than the other, and how do we measure that? Thus, the process of medicalization and the availability of new technologies like the CI are fraught with gains as well as losses. This is *ambivalent medicalization* in a nutshell.

The ambivalent social consequences that occur because of medicalization include gains and losses that occur on broader, societal, and community levels; ambivalence is not limited to the level of individual experience. For example, the institutions and structures in place to identify and intervene upon discovery of hearing loss in children are extensive and occur earlier in a child's life than ever before. This has changed how early deaf children are identified and improved intervention services, but it has also reshaped how families respond to deafness and how the clinical practices surrounding the CI are implemented, and has increased the therapeutic labor required of mothers. Earlier identification is better, no matter how it is responded to. But while medicalization has resulted in deafness being identified earlier, it has also depoliticized it. Much of the political controversy over CIs does not touch these families' lives, precisely because they are surrounded by a host of formalized programs that adhere to a medicalized script. Today, each state runs federally mandated newborn screening programs, and they run these NBHS programs in conjunction with hospitals that refer children to audiologists for follow-up testing. The technologies deployed during this process—such as OAE testing, ABR testing, and CIs—work in tandem with the social technologies in place to surround parents and anticipate their emotional responses to the news. Newborn screeners and audiologists carefully deploy communication strategies. The medicalized script of deafness is both discursively produced by experts in this space and reflective of larger cultural values concerning disability and deafness. That is, mothers and medical professionals bring with them their own *latent scripts* about deafness and disability and thus also contribute to the production of deafness in interactions. In effect, the way implantation is structured today renders controversies over CIs and the arguments against them by the Deaf community largely moot and neutralized.

From a broader sociological viewpoint, the organization of implantation and the development of the CI demonstrate the *possibilities* that medicine offers as well as the *labor* it demands. This is the central tension of ambivalent medicalization. For example, the therapeutic culture surrounding CIs holds the possibility of relief for parents through a promised narrative of overcoming deafness, but it also demands their labor in exchange (not to mention that it will alter the Deaf communities of the future, a point I will take up again later). This labor is not only gendered but directed at a specific site in the body: The redefinition of deafness from a sensory to a neurological problem displaces the responsibility for "success" from the device to the mother, and specifically her ability to "train the brain." Families (primarily women) are responsible for overcoming their child's disability and for engaging in (neuro)scientific motherhood to raise their children. As a result, when mothers are thrust into these moments by a failed NBHS or an ABR diagnosis, they experience a range of conflicting and ambivalent emotions about the situation, such as *anxiety and hope* or *grief and relief*, and about the labor they must perform. And even as power is wielded over the individual mothers in this study, there is also agency in their actions, as the promise of the CI technology offers a chance to participate in a narrative of disability whereby normalcy is restored or achieved (Frank 2013).

The institutionalized and highly structured labor in which parents are expected to engage requires hope. This hope is sustained over a number of years by carefully structured and institutionally formalized practices that address the "soft" aspects of implantation. This includes opportunities to attend support groups and therapy with the clinic's social worker and parent communities, managed or orchestrated by the health care system within which families find themselves and learn to navigate. So while it is demanding and must be surrendered to, medicalization also *generates* new social worlds. Through these social structures, parents create community and share a common experience, coming, as Jane said, "full circle" in giving support to other parents when it is needed. They might have started in grief, but over time (especially if the CI "worked") they share their experiences with others and provide support. Indeed, some even referred to it as "CI culture." And yet, sometimes parents *resist* the imperatives placed on them to restrict their child's exposure to sign language, question their denial of visual language and cues, and rewrite the script of deafness to include aspects of sign in their daily lives.

Medicalization sparks questions about larger *social and ethical* implications that, depending upon your own politics or philosophy, are either good or bad. These implications can play out on a number of levels. Some consequences occur at the level of the individual or family. For example, the anticipatory structures and therapeutic culture of implantation likely *prevent the possibility* of mothers believing that they can raise a healthy deaf child outside the medical gaze. This is a powerful social process that changes the experience of motherhood. For example, mothers who begin from a state of grief at diagnosis are immediately surrounded by structures that anticipate their emotional needs and individuals who share a united message. This message primarily advocates against sign language, even as mothers may struggle to communicate with their children and question their own abilities to maintain their children's safety and understand their desires or needs. Clinical staff and state EI service providers supply emotional support, but they also supply a specific narrative and path. This path is predominantly characterized by an allegiance to the future; interventions now that focus on speech, even if the child is deaf, are framed as paying off later, as ways to ensure independence, success, and normalcy. At best, this process attempts to rigidly structure how mothers interact with their children and, at worst, may undermine mothers' trust in their own instincts and desires to communicate in the present.

The anticipatory structures and therapeutic culture of implantation also socialize mothers to value the goal of vanquishing their child's deafness through biotechnological means—a goal to be achieved through their own labor. This, however, depends on ableism and the goals of normalization, which means upholding the objective of fitting the body to social norms rather than asking society to accommodate differences. On the one hand, this devalues different ways of being and depends on the exclusion of sign language. On the other hand, the project of normalization is rewarded in part because it provides access to a specific set of institutional and therapeutic resources and children with advantages in the future. Throughout this process, hope is fostered because there exists the possibility that the CI will "work," and that the deaf child can eventually pass as hearing with the CI coupled with successful long-term therapeutic labor. In this regard, a child might grow into a deaf adult and into a community of CI users or the hearing world. Nancy, in one interview, talked to me about her daughter Anne's best friends from the annual AGB conference; they all had CIs. This introduces another set of social implications regarding the creation of new identities and communities. Because of the growing numbers of implantees, it is possible that the Deaf community may dwindle as a new CI community emerges. Philosophers and d/Deaf persons alike have questioned the ethics surrounding this possibility, while others promote a "deaf futurism" that embraces ever-expanding ways and experiences of being deaf.[2]

When mothers adhere to medical advice and succeed at auditory training, they speak about their achievement at being a *good mother* because they have done the work. Accordingly, parents such as Nancy believed that the CI is not to blame if a child does not learn to hear and speak. Instead, such failure indicates that the mother has not done enough. That is, a child's success with the CI demonstrates the mother's capability, and, in the biopolitical era where health is a moral obligation, perhaps even her morality. For middle-class mothers in particular, this kind of high-intervention, therapy-oriented mothering fits into their expected parenting style. Thus, for the mothers in this study, their child's success shows whether they have managed to do the work, to meet the therapeutic demands, to master the technology, navigate the institutions, and demand the services. Clinic staff reinforce the value of this social role, labeling those who do not do the work as "difficult moms" and claiming that most "normal" parents, who share a concerted cultivation style of parenting, decide to implant.

Finally, medicalization produces a range of intended and unintended *consequences* and *tensions*. These consequences and tensions may be simultaneously both positive and negative, and, as all of these examples have shown, they span a broad range of social spheres. They may occur at the site of health care delivery, such as the tensions between tending to families' needs and managing their emotions in order to maintain efficiency and optimize implantation outcomes within the day-to-day operations of the clinic. On a broader scope, medicalization and new technologies redefine what "good mothering" looks like in the context of having greater treatment options. More often than not, the pressure to utilize all options available reveals dominant cultural narratives that deem some bodies acceptable—those that can hear and speak—and others not—those that do not hear and that use sign language.

At the same time, individual families experience the benefits from implantation—many of the implanted children I encountered can pass as hearing, and all of the parents are pleased with the outcomes they have experienced so far. This provides relief to parents whose children are doing well in spoken language development. As Julia and Paul said of Morgan, they have every expectation that he will be an independent, functioning adult. In middle-class American families, this is one of the most dominant narratives of what successful parenting looks like, and thus they feel validated in their choices and proud of their accomplishments. For the most part, however, parents are unaware of being socialized, and professionals are unaware of operating from their own value systems. The medicalization of deafness is largely taken for granted as common sense. But in this book, I have pulled apart these moments in the clinic, in the school, and in families' lives. Slowing down interactions and subjecting them to sociological analysis reveal that middle-class hearing parents share with the medical establishment an ethos of parenting and a set of values about disability and language. This allows for a seamless overlap of framing their child's deafness and carrying out the task of implantation. What effect might this have on future d/Deaf communities? Adherence to implantation techniques and fulfillment of labor imperatives by mostly white, middle-class families may reproduce social inequalities and leave intact normative narratives of disability and mothering.

Reproducing Inequalities

Class and Cultural Background

Disparities and a middle-class bias in implantation have been identified for some time (e.g., Luterman 1991). Rates of implantation for white children and Asian children have been previously documented as five times higher than Latino/a children, and ten times higher than black children; similarly, rates of implantation in areas with above-median income are higher than those in areas with below-median income (Hyde and Power 2006; Holden-Pitt 2000). These same disparities in implantation have been observed in the United Kingdom as well (Fortnum, Marshall, and Summerfield 2002). None of these studies has identified precise causes for these differences, although Chang et al. (2010) later conducted a study that found no disparities in *access* to CIs but rather in the subsequent long-term follow-up care and therapeutic labor required. Boss et al. (2011) present similar patterns in class disparities but conclude with a call to more clearly identify and understand this process.

I propose that these disparities are not only structural but also cultural. In this book, I have relied on previous research by Lareau (2003) that describes how parenting styles and parental relationships with institutions differ across class statuses. In the preceding chapters, I have shown the importance of the long-term feature of implantation; it is a complex, ongoing, and institutionally embedded process. The labor demands characteristic of the therapeutic culture of implantation, the resources needed to navigate the various cooperating institutions, and the high-intervention style of parenting required most closely align with the middle-class parenting style that Lareau identifies as concerted cultivation. While the uniformity and size of the sample for this study preclude the ability to draw conclusions across categories, the qualitative description of the social organization of implantation gleaned here nevertheless suggests that the values of concerted cultivation permeate the various systems that serve deaf children today; that is, the downstream disparities are caused not just by structural variables but also by cultural ones.

Such disparities indicate that the medicalization of deafness and the advent of the CI do not result in "solving" deafness. Rather, the technological fix of the CI results in reproducing preexisting inequalities. Thus, understanding disparities in implantation is important for improving outcomes across demographic groups. But it is also important to acknowledge that CIs are not a panacea and that they can exacerbate inequalities within deaf communities. Because of disparities in implantation, a new deaf underclass—marked by the use of sign language *or* underdeveloped language—may emerge in the future.

The results of my study and others suggest that because of the demanding nature of implantation, specific groups of deaf persons may emerge as particularly disadvantaged in the future: poorer, nonwhite children for whom the CI "fails" because of inadequate resources (e.g., economic or rural/urban divides) or an inability to comply with long-term therapeutic labors (especially if sign language is discouraged from the outset, then they may end up without any language at all, which creates additional developmental problems) and immigrant families whose children do not receive CIs or receive poor follow-up care. As I also showed in this book, these disadvantages occur not only because of differences in parenting styles but also because of biases that professionals in implantation may have against non-middle-class parenting styles, non-English-speaking families, and Deaf families.

Ability/Disability

Another social category of inequality is disability. Disability studies scholars have long showed how rather than being strictly a medical or bodily characteristic, disability is also a vector of social discrimination and a social category of difference. Moser wrote that normalization efforts directed toward those with disabilities are "constantly counteracted by processes that systematically produce inequality and reproduce exclusions" (2000, 201). Throughout this book, I have shown how the promise of CI technology for parents is tethered to the possibility of "overcoming" their child's deafness. That is, medical technologies such as the CI are often meant to transform disabled bodies into "competent normal subjects" (Moser 2006, 375). I have also shown how "success" in the context of implantation is dependent on the ability to transform the brain. Sign language users are often the "failures" relegated to the Deaf world. Thus, a technology meant to equalize instead reinforces social hierarchies and reconfirms and reproduces the very categories it seeks to dissolve.

The Neuroscientific Reconstruction of Deafness

One of the most salient features of implantation is the centrality of the brain. Neuroscientific discourse influences everything about how we think about deafness: where the "problem" of deafness is located, what gets researched, how we set up schools, what mothers are expected to do, and how communities on either side of the scripts of deafness see each other. Neural narratives about deafness directly map onto and reinforce social divides. Neuroplasticity is characterized in specific, if inaccurate, ways in order to accomplish this: On the one hand, the brain is passive and must be trained by parents, and on the other, it can also be unruly and out of control. The latter claim is made particularly in relation to sign language exposure; if a child is allowed to sign, then his or her visual pathways can form, preventing auditory pathways from forming.

The discourse on the brain and language in implantation is untenable. Because science accomplishes cultural work, it is unsurprising for one side of a debate to make scientific claims that support its own position. It is also not unusual for communities of scientists to pay attention to some research and not others. While I am critical of the use of neuroscientific discourse to achieve cultural ends, here I turn to the possibilities of using some counterneurological arguments precisely because they do have power.

It is a commonly held belief that early childhood exposure to two different languages is detrimental. As I have shown over the previous chapters, this belief is especially powerful in the context of bimodal (sign and speech) bilingualism, deaf children, and implantation. It is so pervasive that it structures clinical practices, EI practices and therapies, and school systems that serve deaf children. But research on bilingualism has shown for more than three decades that the best time for language acquisition is birth to age five and that "if a child has two languages exposed to them during that period, rather than being confused or delayed, the human brain takes the neurotissue that has been biologically endowed for processing language and grabs it and sets it up and begins to build these pathways in two discrete systems" (Petitto 2014).

In perhaps the greatest irony, trends in using Baby Sign Language (www.babysignlanguage.com) and other similar programs claim that teaching your hearing baby sign language will boost his or her overall language development. And yet the clinical ideology about exposing deaf children to sign is so powerful that it is common practice to withhold sign language from deaf children. But recent studies "tested the hypothesis that if a young child has cochlear implantation that you should not expose a child to sign language because . . . it will hurt auditory tissue development" (Petitto 2014). Results showed that implantees who received early exposure to sign language had entirely normal auditory tissue development. In other words, "sign language exposure did not hurt the development of auditory tissue or . . . cause it to be deviant in any way" (Petitto 2014). Furthermore, other linguists have found that bilingualism—the norm in most of the world outside the United States—is beneficial, which appears to hold true in the case of bimodal (visual and auditory) bilingualism in CI recipients (Davidson, Lillo-Martin, and Chen Pichler 2014; Lyness et al. 2013; Hassanzadeh 2012). But literature written in fields associated with implantation, such as otolaryngology and audiology, argues quite the opposite about bimodal bilingualism in children with CIs.[3]

As it stands, the current version of neuroplasticity in implantation results in children without easy access to language until surgery (which can be a year or more), contributes to the political divisions between language choice and communities, and maintains the controversy over their use. Relying on such neural narratives also demonstrates our cultural affinity for reducing the social complexities of deafness (and many other things for that matter) to the firing of neurons. Even as I have been critical of the way professionals in implantation have deployed neuroscientific knowledge to achieve cultural ends, it is advantageous and crucial for those advocating sign language exposure to begin or continue to deploy neuroscientific claims of their own, which could provide further opportunities for deepening the debates. And this is a crucial task, as the separation between the Deaf and CI communities is also untenable.

What Questions Should We Be Asking?

An era in which bodies are increasingly able to be reshaped, brain architecture specifically worked upon, and disabilities mitigated through medical intervention produces important new questions that should be asked. For example, how can a consciousness about the experience, culture, and politics of disability be included? In the specific case of raising children with disabilities, Gail Landsman's (2008) study of mothers of children with developmental disabilities showed that they engaged with both the social and the medical model of disability. Feminist disability studies scholar Alison Kafer writes that her political/relational model of disability "neither opposes nor valorizes medical intervention; rather than simply take such intervention for granted, it recognizes instead that medical representations, diagnoses, and treatments of bodily variation are imbued with ideological biases about what constitutes normalcy and deviance. . . . it recognizes the possibility of simultaneously desiring to be cured of chronic pain and to be identified and allied with disabled people" (2013, 6). This model of disability most aptly fits with the ambivalence of medicalization that I have argued here, one that leaves room for both the good and the bad consequences of medicalization and removes the necessity of making stark choices, for example, between scripts of deafness and models of disability.

In the case of deafness, hearing parents are the primary consumers of CIs, and they will continue to implant their children. First, then, the imperative here must be to figure out a way to assist those with disabilities who are using medical technologies in living well with them while not making them feel bad about themselves.[4] Thus, we might ask, what might be gained by implanting children *and* assuring them that their deafness is acceptable and even worth celebrating by exposing them to the possibilities of sign language and the values of the Deaf community? What might be gained from embracing mothers' choice to implant *and* inviting them into the Deaf community?

Second, mothers struggle to communicate with their children before they begin to acquire spoken language, when the CI is unavailable or turned off, and when there is a safety concern and sign is the most expedient mode of communication. But the therapeutic culture of implantation described in this book is often hostile to sign language. Some questions to ask about this include the following: Is the monolingualism that the CI community advocates a best practice? Would an end to the anti–sign language rhetoric decrease disparities in language development? Would it ease the stress that mothers feel when trying to communicate? What benefits would this have? Furthermore, how might best practices be defined when there is a lack of adequate data on implantation outcomes and even a comprehensive understanding of what outcomes might be measured?[5]

Kermit (2012) makes a compelling case that there is a continuum of the Deaf critique and that professionals should heed the concerns that appear in what he terms the moderate Deaf critique in order to answer such questions.[6] For example, in considering the usefulness of some Deaf persons' concerns and explaining the moderate critique, he writes, "It is not the cochlear technology in itself they view as problematic, but rather the subsequent rehabilitation process. Because they themselves have experienced what they describe as harmful effects, which relate above all to the idea of normalization, they have articulated worries for the new generations of deaf children in need of rehabilitation following cochlear implant surgery" (Kermit 2012, 367). In taking such a view, the question to ask, then, is less about evaluating whether getting a CI is a good choice, and more about asking what the possible consequences and outcomes of implantation might be. Part of such a moderate critique understands the desire for spoken language but also "[doubts] that 'spoken language' is a satisfactory measure for evaluating the ultimate outcome of implantation" (Kermit 2012, 368).

I have two suggestions for a more comprehensive evaluation based on a moderate Deaf critique and the results of the research presented in this book The first suggestion is that there should be a holistic focus on language and authenticity that involves recognizing both speech and sign as equal and acceptable modes of communication. This would lessen pressures on mothers, as well as indicate to children that whatever form their communication takes is acceptable. It would also allow children to have immediate access to linguistic input, thus eliminating the language deprivation that may occur between diagnosis and the auditory training that begins after implantation. This need not be seen in opposition to preparing for and receiving a CI. The second suggestion is longitudinal: This study did not focus on the experiences of children with CIs, but as the cohort of those who are experiencing these relatively new interventions grows into young adulthood, researchers should gather their stories. How do children integrate this level of medicalization from such a young age and see themselves as a result of it? To date, only a handful of studies have attempted to gather such data. In the past, the common narrative of Deaf persons' experience was of being without any language at all until they found ASL and were exposed to the Deaf community. But today, the narrative of deaf childhood is different. Two small studies examined the experiences of children with CIs, one in the United Kingdom (Wheeler et al. 2007) and the other in Sweden (Preisler, Tvingstedt, and Ahlstrom 2005). In the former, some of the children talk about using sign with deaf people and speech with hearing people. In the latter, they all identify as bilingual and bicultural. But as Fjord's (2001) previous comparative study of the culture of CI clinics showed, the United States is particularly insistent about monolingualism in English, while Scandinavian countries are not. What we do not know yet is how the CI technology influences how children view themselves and the identities and communities that they grow into.

I ask these questions from a position that advocates at least the possibility of using sign language with deaf children, regardless of implantation status. Implantation and a familiarity with sign do not have to be separated. As seen in my study, ideology keeps them separate now, but the audiologist David Luterman, who has over fifty years of experience with deaf children, states that "a variety of approaches are needed . . . the child will tell us which is the best way for him or her to be taught. More research needs to be conducted so we can match the child to the methodology sooner. At the present time it is more trial and error" (2004, 18). He goes on to validly criticize the Deaf critique of implantation, stating that it is out of step with larger cultural norms and that hearing parents are significantly impaired in their ability to be language models for their children in ASL since it is a language they do not know. These are valid points but not insurmountable obstacles. Better access to sign could provide parents and children a means of communication no matter where they are in the CI process, reduce mothers' stress and increase parent-child bonding, connect more individuals across communities, mitigate spoken language acquisition disparities, and engender more nuanced views about the role of CIs in society and in the Deaf community. This is not, like the implantation-related rhetoric, a plea for monolingualism but rather a position that advocates bilingualism in implanted children. However, it should be noted that I do not see bilingualism as a panacea for disparities in implantation; various other social spheres would need to be addressed. For example, Knoors and Marshark (2012) give a nuanced discussion of findings around bilingual/bicultural education, and they also cite the difficulty of having hearing parents be a language model for their children when they themselves do not know sign language. Nevertheless, they argue, the benefits of access to sign for deaf children are myriad, especially in the sphere of psychosocial development. In other words, various social changes would have to occur to address the complexity of these issues, and cooperation across the divides over implantation would be necessary. Such an integrative approach to implanted children would also require change on the part of those on both sides of the debate. And while this book has focused on audiological practice, this approach would also need to include the extreme ends of the Deaf critique of implantation and would require the Deaf community to be more open to individuals who use CIs, as well as to the parents who chose them for their children. The key to the survival of the Deaf community is in realizing the many different ways one can be d/Deaf.

Critical Alliances

One of the first things I noticed in doing fieldwork at a CI clinic and with families who have an implanted child is the lack of overlap with the Deaf community. Given the history of the fierce debates over the CI, this is unsurprising. This divide should be dismantled; as Bauman (2008) states, the division between sign language and CIs is a false dichotomy. Nevertheless, the empirical data from fieldwork, albeit from one clinic, demonstrate that it is a powerful and entrenched divide. And yet over the years, I have given numerous talks on this subject in different cities, and each time professionals who work in CI clinics or are educators have told me about how different their clinic, school, or community is from the one I studied. This indicates the dire need for (1) a greater understanding of how clinics across the country are structured and what their therapeutic cultures look like, and (2) a more uniform set of recommended practices based on such research, a need also highlighted by leaders in implantation (e.g., Sorkin 2013). Those who adhere to the Deaf moderate critique and professionals and researchers in implantation may find some common ground in such endeavors.

Another critical alliance could be formed between the Deaf community and mothers of deaf children with CIs. Both communities should pay attention to women's stories, because, as this book has shown, it is the mothers of deaf children with CIs who must live out medical, clinical, and educational policies that get made. For example, some mothers have shared with me that there are vast online communities of those with deaf children who have CIs, and the women in these communities are innovative and adept at living in the gray areas between the dichotomous debates that play out in the public sphere. Their voices were not part of this study, but what would happen if the covert bilingualism that some mothers encourage with their implanted children was accepted and studied further? Given the rates of implantation, the increasing number of children who have CIs, and the diminishing amounts of resources going to traditional sites of Deaf cultural production (such as Deaf residential schools), these families are some of the emergent sites for Deaf cultural production. These are the critical alliances that will need to be built in the future.

Families and Social Structure

When a child is born today, protocols and institutions are in place to immediately evaluate his or her health. Mothers most likely give birth in a hospital system, most likely with every expectation that their child will be able to hear. When a child is found to be deaf, however, families are surrounded with a set of anticipatory structures that funnel mothers through a set of steps. The data show that if the child meets the audiological *and* social criteria for a CI, he or she will most likely receive one. Today, more parents are choosing implantation for their deaf children than ever before.

Processes of medicalization are so pervasive that disability, and specifically deafness, is made in relation to technology and scientific knowledge. But families of different classes or cultural backgrounds encounter these systems differently. Mothers are increasingly therapeutic laborers; as technological interventions increase, so do the labor demands. And in the era of the brain, mothers understand their duties in neuroscientific terms. The site of their labors has gone deeper into the body, down to the wiring of the brain. It turns out that medicalization is not "all good," nor is it "all bad." Rather, it is ambivalent; demanding and rewarding, and with limitations and possibilities. However, the shape that these possibilities and limitations take is influenced by social factors. This shifts the discussion from whether the CI is a good or a bad technology to instead examining how it is implemented and who benefits. By gaining a better understanding of these processes, such as of the structural and cultural aspects of implantation that I have described in this book, we may begin to more adequately address deafness in children today.

ACKNOWLEDGMENTS

I waited a long time to write this section of the book, and it gives me great joy to thank the many people who made *Made to Hear* possible. First and foremost, I am deeply indebted to the individuals who participated in the research. In the interest of maintaining confidentiality and anonymity, I cannot thank each of you personally here, but I am acutely aware of the gift I was given by being allowed to write these stories, especially as I sought to write about an issue as controversial as this one. I particularly thank the study participant known as Jane in the book: your openness and willingness to be emotionally authentic with me made this work what it is, and I hope sharing your story will give other parents comfort as they see their own experience in yours.

This book is based on my dissertation fieldwork while a graduate student at the City University of New York, The Graduate Center. I am grateful to the members of my dissertation committee: Barbara Katz Rothman, Victoria Pitts-Taylor, and William Kornblum. I especially thank Barbara for her mentorship and wisdom in academic, as well as personal, endeavors. While a graduate student, I was privileged enough to be mentored at The Hastings Center and Yale University's Interdisciplinary Center for Bioethics. I thank William J. Peace for referring me to the Yale–Hastings Program in Ethics and Health Policy; it undoubtedly improved my work. At The Hastings Center, I especially thank Erik Parens for his guidance, thoughtfulness, and generosity of spirit. At Yale, I thank Steve Latham for his continued encouragement and enthusiasm about the project, and Carol Pollard for her assistance.

Parts of this book have been published over the years as articles, and the peer review process contributed greatly to the level of analysis later developed in this book. I thank the anonymous reviewers at *Sociology of Health and Illness* and *Science, Technology, and Human Values.* I began writing the initial draft of this book during my year as a visiting professor at the Rochester Institute of Technology. I thank the community of scholars at RIT who helped me get started, especially Deb Blizzard for her unbelievable support and mentorship, as well as John Albertini, Catherine Clark, Carol De Filippo, Sue Foster, Jennifer Gravitz, Pamela Kincheloe, Marc Marschark, Ila Parasnis, Vince Samar, and Michael Stein. You each gave me your time or shared your expertise, and some of you even commented on drafts of chapters.

I am deeply indebted to my editor, Jason Weidemann, at the University of Minnesota Press. I thank him not just for his extraordinary guidance and thoughtful feedback but also for his early interest in my work. I benefited greatly from his support and vision, as well as from numerous others at the University of Minnesota Press and the Institute for Advanced Study, including Anne Carter, who had faith in my ability to make the work better and so patiently provided feedback on early chapters of the book. I also thank the reviewers of the book manuscript, who provided in-depth criticisms and pushed me to make this work more than I ever thought it could be.

The final stage of manuscript preparation was supported by the University of Connecticut Faculty Small Grant program. This allowed me to work with Chaia Milstein, whom I wish to thank for her thoughtful and careful editing of numerous drafts. I also benefited from feedback from Marie Coppola, Matt Hall, Diane Lillo-Martin, and many others in the Sign Language Research Discussion Group at UConn.

I thank Brad Gibson for opening up the Deaf world to me. Without our friendship when we were just kids, the possibility of this book would not even exist. I still remember the day in the gym at school when you had to spell *hurricane* out to me numerous times before I finally understood what you were saying. Thank you for teaching me sign language and shaping the trajectory of my entire life. Thank you, Byron Bridges, for sending me to Gallaudet University in the first place, for telling me I could do more, and for giving me the confidence to do it. Thank you, Dirksen Bauman, for mentoring me not only while I was at Gallaudet but also during the years beyond. And a heartfelt thank you to the Deaf communities far and wide—from Austin, Texas, to Washington, D.C., Bristol, England, and New York City.

Numerous other colleagues, mentors, and friends contributed in both small and large ways over the years. Some are longtime friends, some simply helped me with a citation when I needed it or let me work out an idea with them, some were colleagues and mentors whom I have deeply appreciated over the years. I wish I had a million pages to devote to thank you each personally: Ben Bahan, Sharon Baker, Anne Balay, Sharon Barnartt, Douglas Baynton, Pamela Block, Stuart Blume, C. Ray Borck, Crystal Bradley, Allison Carey, Lawrence Carter-Long, Monica Casper, Christina Castillo, Michael Chorost, Dána-Ain Davis, Jeanette Di Bernardo Jones, Katherine Dunn, Alice Eisenberg, Michele Freidner, Vasilis Galis, Shira Grabelsky, Ed Hackett, Aimi Hamraie, Joseph Hill, Andrew Hoffman, Thomas Horejes, Kathryn Housewright, Robert Hughes, Emilio Insolera, Alessandra Lacorazza, Greta LaFleur, Andy Lampert, Gail Landsman, Riva Lehrer, Valerie Leiter, David Linton, Simi Linton, the Mate family, Cricket McLeod, Michael Ian Miller, Mara Mills, Joseph Murray, Karen Nakamura, Nancy Naples, Paul Pomerantz, Michael Rembis, Sal Restivo, David Roche, Harilyn Rousso, Claudia Secor-Watkins, Richard Scotch, Cynthia Schairer, David Schleifer, Amy Sharp, Hannah Sheehan, Melinda Shopsin, Fern Silva, Wendy Simonds, Adrienne Sneed, Sel Staley, Bethany Stevens, Joe Stramondo, Christine Sun Kim, Sunaura Taylor, Gregor Wolbring, Anna Wolk, Sophia Wong, and especially my many comrades at the Society for Disability Studies and in the New York City disability rights scene.

Thank you to amazingly supportive members of my family: Suzan Brown, CJ Dowling, Jane Dowling, Tristan Marsh, Bradley Mauldin, Dan Mauldin, Stan Mauldin, and Lois Murphy. I also wish to acknowledge Ameerah, my precious and dearly departed dog of nearly seventeen years who almost lived to see the book that kept me at my desk and delayed so many of her walks. And lastly, I want to acknowledge possibly the most important and far-reaching influence on my life and my work. My approach toward the research in this book was deeply shaped by the love and loss of my partner, Jamie Finkelstein. Though she died before I even finished the dissertation, the lessons that came with becoming her caregiver made me a better sociologist and a better person.

NOTES

Introduction

1. All names have been changed to protect the identity of the study participants.

2. The word *deaf* defines an audiological state, while *Deaf* indicates a cultural identity stemming from this state. Being "Deaf" is an identity primarily characterized by one's use of American Sign Language (ASL) and embodiment of a set of distinct cultural values (Padden and Humphries 1990). Throughout this book, I use deaf and Deaf where appropriate. To be inclusive of both terms, or when referring to their coexistence, I use "d/Deaf."

3. When the dialogue in this book is placed in quotation marks, it was taken directly from interviews and observations that were recorded on a digital voice recorder. However, not all interactions were recorded and directly transcribed. Sometimes I did not have permission to record. In those instances, I reconstruct the dialogue—without using any quotations—from the extensive notes I took during the interactions. Thus, if there are no quotation marks around dialogue or if quotation marks appear only around a very specific phrase that I wrote down word for word, then it was taken from my field notes and not a recorder.

4. There is much literature on outcome predictors in pediatric cochlear implantation, although results are still variable. Nevertheless, my focus is on the parents' experience of this unpredictability because implantation is not equivalent with immediate spoken language acquisition.

5. Disability is a broad and wide-ranging category and thus an umbrella term. According to the World Health Organization (WHO), disability is a complex social and embodied phenomenon, one of whose characteristics is that activities of daily life are limited (WHO 2014). The WHO categorizes deafness as a "sensory disability," as does the Americans with Disabilities Act (ADA 1990) and the United Nations Convention on the Rights of Persons with Disabilities.

6. For some examples of works that outline the Deaf cultural perspective and critique of culture, see Bauman's (2008) edited collection showcasing work in Deaf studies, Bragg's edited collection (2001), the now classic

texts by Padden and Humphries (1990) and Lane (1989), and Ladd's deeply theoretical work (2003).

7. Campbell (2009) also provides an excellent critical analysis of the marketing of cochlear implants to families in a chapter of her book *Contours of Ableism.*

8. Many studies of decision making have been undertaken, including Nikolopoulos et al. (2001), Christiansen and Leigh (2002), Li, Bain, and Steinberg (2004), and Okubo, Takahashi, and Kai (2008). Christiansen and Leigh's (2002) excellent book *Cochlear Implants in Children: Ethics and Choices* is based on interviews with parents about numerous other aspects related to pediatric implantation, such as decision making, variability in outcomes, language development, and education.

9. See Callon, Law, and Rip (1986), Kuhn (1996), Latour (1988), Sismondo (2008), Law and Mol (2002), and Winner (1980) for examples.

10. The larger questions then are these: Given the medical technologies available to us today, to what extent should we use them to make our children mirror the norms of society? Should we embrace differences, including disabilities, or "correct" them? In his edited collection *Surgically Shaping Children*, Erik Parens (2008) brings together scholars from a variety of fields, including disability studies scholars, bioethicists, parents, and people with disabilities. What emerges from this collection of essays is the idea that once the scholars examine each individual condition in depth, their judgments about whether to medically intervene vary for complex reasons. Certainly, we have only begun to encounter and consider the ethical problems and social implications of powerful medical technologies, and it is outside the scope of this book to try and resolve them. However, I follow Parens's (2008) strategy of going about this task of thinking through the larger questions by narrowing the scope. In this way, by examining the specific technology of CIs, the specific condition of deafness, and the interactions and practices taking place in a CI clinic between professionals and parents, we can perhaps chip away at these larger questions.

11. A number of other issues, such as deinstitutionalization, community participation and integration, education, work, transportation, and housing, emerged from the Independent Living Movement, as did a general demand for less stigmatizing social attitudes (Shapiro 1994). As a result of this movement, the field of disability studies emerged and focused on the political aspects of disability in the 1970s in both the United States and the United Kingdom (which also currently has an active disability rights agenda and movement). See the National Council on Independent Living website at http://www.ncil.org for a detailed history and philosophy of the move-

ment and its social justice work, especially around deinstitutionalization. Also, see the grassroots U.S. organization ADAPT (http://www.adapt.org).

12. These ideas have been articulated across disciplines, such as in social science, memoirs, women's studies/feminist theory, and disability studies. A brief sample here from these disciplines includes Clare (1999), Linton (2007, 1998), Mairs (1997), Saxton (2000), Wood (2014), Kafer (2013), L. Davis (2013), Hall (2011), Barton (1996), Corker and French (1999), and Oliver (2013, 1990).

13. This separation of deafness and disability and the articulation of a Deaf-specific model can be seen in texts such as Lane (2005), Eckert (2010), Hauser et al. (2010), Reagan (2002), and Ladd (2003).

14. While I do not take up a neurodiversity approach explicitly in this book, it could nevertheless be extremely relevant in future discussions of implantation since, as I detail in chapter 4, the brain is the primary site of "work" in implantation. See Harmon (2004), Savarese and Savarese (2009), and Sinclair (1993) for further reading.

15. There is much scholarship on the intersections of medical or technological capabilities and the implications for disability, for example, Rothman (1993), Saxton (2000), Colligan (2004), Rapp (2010, 2000), Landsman (2008), Moser (2006), Parens (2008, 2000), Parens and Asch (2000), and Scully (2008).

16. Increasingly, scholarship in disability studies is taking on the limitations of the social model (e.g., Shakespeare and Watson 2001; Shakespeare 2006; Siebers 2008; Kafer 2013).

17. Shakespeare and Watson (2001) critiqued the social model for being too naive. Shakespeare (2006) later developed what he calls a "critical realist" model of disability, while Siebers (2008) developed the concept of "complex embodiment." Both of these models tried to integrate the realities of embodiment and give the body some role in the experience of disability. While the social model has been profoundly important, Shakespeare's critical realist model is more resonant for me, where "impairment is not the end of the world, tragic and pathological. But neither is it irrelevant, or just another difference" (2006, 62). The social model tends to deny embodiment to the point where if you study aspects related to medicine, then your work would be relegated to the medical model. There were some unintended consequences in disability studies because of this split between the two models: (1) the realm of the medical was somehow construed as *not* social, and (2) like early feminist theory, the focus on political struggles was predicated on theorizing the body away. Perhaps alarmingly to some, I have focused on medicine in this book, considered parents' accounts of their child's marked body, and investigated what it means to care for them. I want

to understand disability by turning a sociological and ethnographic eye to bodies and their caregivers in clinic and home and to the very practices that are widely viewed as "medical." I am not advocating that medicine should be used to alter individual bodies but rather pointing out that (1) there is a difference between situating bodies as *deterministic* and acknowledging them as *influential*, and (2) the medical and scientific realm *is social*.

18. An excellent edited collection on the topic is Lewiecki-Wilson and Cellio's (2011) book *Disability and Mothering*.

19. Fjord's comparative study of CI clinics (1999 and 2001) showed how the scripts around implantation included the brain and are organized differently in different cultural contexts, providing useful cross-cultural models for understanding how implantation practices differ across the world. Such studies deserve greater attention and development.

20. See Moeller (2000), Belzner and Seal (2009), Chang et al. (2010), Stern et al. (2005), Conger, Conger, and Martin (2010), and Yoshinaga-Itano (2003b) for further reading.

21. I want to draw attention here to the difficulty of doing ethnographic fieldwork in a clinical setting. Sociologists Anspach and Mizrachi argue that medical ethnographers push the ethical limits of ethnography and note that the relationship between the academic fields of medicine and sociology is antagonistic, asserting that sociologists are "perennial irritants to those we study" (2006, 714). The well-known medical sociologist and ethnographer Charles Bosk also writes that "we betray our subjects twice: first, when we manipulate our relationship with subjects to generate data and then again when we retire to our desks to transform experience to text" (2001, 206).

1. A Diagnosis of Deafness

1. Hearing aids are generally not covered by insurance, but as I will discuss later, CIs are. According to the American Speech-Language-Hearing Association, hearing aids are sometimes covered by private health plans but not usually in full. They are, however, often covered for children through Early Intervention programs. There are also tax credits available for them. In general, hearing aids are not covered, although audiologic testing and evaluation for them is. For more information, see http://www.asha.org/public/hearing/Health-Insurance-Coverage-for-Hearing-Aids.

2. I also want to acknowledge Kathryn P. Meadow's early sociological work (1968) on this topic where she also showed that deafness was "invisible" until much later in the child's life and described the effects that this had on parents.

3. At the request of Margaret, a digital voice recorder was not used

during interactions with her. Therefore, all of the reported dialogue with this subject was taken from the extensive notes I took during our conversations and not directly from recordings. Any errors in reporting what was said are my own.

2. Early Intervention

1. In a status report for the state of Connecticut, almost 80 percent of children diagnosed with a hearing loss through NBHS were subsequently enrolled in EI services (Honigfeld, Balch, and Gionet 2011). New York State (NYS) specific data are thus far unavailable. However, the NYS Department of Health released a report stating that with CDC funding they will now begin to collect data: "This will be the first time that NYS has collected individualized data on newborn hearing screening results as well as results on follow-up for those infants who do not pass their newborn screening. Both the EHDI information system and the early intervention system contain fields that will facilitate linkage and allow analysis to decrease the loss of follow-up of infants with suspected hearing loss" (NYS Department of Health Division of Family Health Bureau of Early Intervention 2013).

2. In the late nineteenth century, Alexander Graham Bell pioneered alternative methods of education that focused on teaching the deaf to hear and speak. Today, the Alexander Graham Bell Association (AGB) is the preeminent national organization for people interested in the oral education of the deaf. AGB has numerous local chapters in each state, providing resources, conferences, and social networks. See their website, http://www.listeningandspokenlanguage.org, for more information.

3. Parents also spoke of extensively using online groups. One of the most popular listservs, with thousands of members, is the Yahoo group called CI Circle. Other websites, like http://www.mymagicfairy.com and http://www.cochlearimplantonline.com, are also popular. There are also numerous Facebook groups for parents of children with CIs.

3. Candidates for Implantation

1. The minimum income reported by families in my study is $60,000 a year, and the maximum income reported is $120,000 a year.

2. Based on data available, white children from higher SES backgrounds who are approximately twelve months of age are implanted at the highest rate. Conversely, EI demographics (across all disabilities) indicate exactly the opposite trend. There is an "overrepresentation of low-income children among EI recipients" in general (NEILS Report 2007, 2–6).

3. Parents and audiologists often talk about auditory memory in a way that suggests that a child who once had hearing and processed information auditorially would retain this ability despite hearing loss, and be better equipped to "decode" the signals the CI provides.

4. Receiving implants in both ears is called bilateral implantation. This practice is becoming increasingly common, especially as bilateral implantation has been documented to produce better spoken language results (e.g., Lovett et al. 2010; Tait et al. 2010; Boons et al. 2012).

5. This lack of evidence-based or outcomes data–based practice should clearly be questioned and critiqued. Part of the work here then is to show what practices are being undertaken, how they are being undertaken, and to what extent claims are made even in the context of "highly variable" and unknown outcomes.

4. The Neural Project

1. So, is this where deafness can be "found"? Feminist theory has long grappled with such questions of whether difference is biologically located. In her work on feminist inquiry and neurobiology, Elizabeth Wilson asserts that feminist scholarship—in further parallel with disability studies—"relies on theories of social construction; in defiance of biological models" (2004, 13). I contend, like Wilson suggests of feminist theory, that disability and Deaf studies can be "deeply and happily complicit with biological explanation" (Wilson 2004, 14). We should be open to neurobiological data in particular, which "need not be at the expense of critical innovation or political efficacy" (Wilson 2004, 16). Neurological discourse is so prominent in this study—brains, neurons, neural pathways, circuitry—and it raises the specter of how social organization flows *from* our understandings of and narratives about the neurobiological, just as much as our socialization and politics influence our understandings of biology.

2. NRT refers to "Neural Response Telemetry," a method of testing that electrical signals are reaching the auditory nerve.

3. Interestingly here, the CI does not only hold maps, but in a sense it also spatially maps the contexts the users enter. A square room bounces frequencies in certain ways, a round room in another. It is a dynamic between the body, prosthesis, and space.

4. They also often get creative in getting around some of the CI's limitations, namely, that it cannot get wet. Many parents mentioned popular YouTube videos that taught them how to use vacuum sealers meant for food storage as a means of encasing the CI in airtight plastic so that a child could wear it while swimming.

5. Throughout this manuscript, participants use the term *oral,* which in this context refers specifically to spoken language; that is, the oral information refers to spoken language input and output. The use of oral in this manner is consistent with a long-established language about "oral education" of the deaf, which refers to an approach to communicating with deaf children exclusively through speech (Baynton 1998). Thus, children are labeled as "oral" if they use speech or are in a speech-centered program. This term is often paired with *auditory* (i.e., "auditory/oral"), which refers to the aural aspects in this educational and communicative approach.

6. Auditory processing disorders can occur in children with normal hearing abilities and be treated through auditory training. See the American Speech-Language-Hearing Association at http://www.asha.org.

7. These "failures" are difficult to access; they may not be in the clinic for regular mapping and were not referred to me, nor would they be immediately identifiable (interventions take years to assess). This points to a hidden population in need of being understood more deeply.

8. The Ling-6 sounds "represent various different speech sounds from low to high pitch (frequency)." According to the Cochlear Americas website, they help to test your child's hearing and check that they "have access to the full range of speech sounds necessary for learning language" (cochlear.com).

9. Hyde and Punch (2011) also found that mothers used sign "unofficially" with their implanted child both for the practical reasons of enhancing communication and the social reasons of introducing their children to the Deaf community and giving them the opportunity to develop a Deaf identity even though they were implanted.

10. The use of sign in the home does not necessarily mean they had mastered sign language. I never observed anyone signing with a child, so I am unable to comment on the level of fluency they had. However, I want to draw a careful distinction here between knowing ASL and knowing a few signs that can be used for a limited number of specific concepts.

11. Focusing strictly on neurobiological characteristics to predict CI outcomes would be a mistake; as I have shown repeatedly throughout this book, social factors may be far more predictive than neurobiological ones.

5. Sound in School

1. In some cases, deaf children at residential schools were punished for using sign language both in and out of the classroom. For striking personal accounts of this, see Cyrus et al. (2005) and Oliva (2004). For a more historical account, see Lane (1989) and Longmore (2003). Wrigley (1997), Lane

(1995), and especially Baynton (1998) also show how Deaf persons and their "natural language" came to be seen as similar to the massive waves of immigrants entering the country at the time, a notion that linked the projects of colonialism and the marginalization of deaf persons.

2. See Jones and Ewing (2002) for an overview of deaf education programs in the United States.

Conclusion

1. Not *all* professionals in implantation overstate the CI's capabilities. For example, Hyde and Power (2006) caution overstating the device's capabilities and clarify that the CI does not make a deaf person becoming hearing. Furthermore, they also acknowledge the effects that rising rates of implantation will have on the Deaf community and point to the possibilities of having it "both ways"—that is, being able to communicate with both hearing and deaf persons—if one is implanted.

2. Ethical and bioethical consequences have most recently been astutely articulated by Teresa Blankmeyer Burke (2006) and Jackie Leach Scully (2008). But others (e.g., Sparrow 2005; Hintermair and Albertini 2005; Lane and Bahan 1998; Ladd and Lane 2013) have also explored the ethical and political questions regarding Deaf communities that changing technologies raise. Meanwhile, Mills (2012) outlines Chorost's (2006) vision of a more transhumanist, deaf futurism, and Friedner (2010) and Friedner and Helmreich (2012) speak to the possibilities for new d/Deaf biosocialities that are possible vis-à-vis implantation.

3. More recently, studies have shown the results of model bilingual programs and put forward new models for seeing deaf children in the context of families (e.g., Snoddon and Underwood 2014; Snoddon 2012).

4. Thanks to Harilyn Rousso for her helpful discussion with me on this question and for the tremendous memoir (2013) that sparked my questions on this topic.

5. Some research, such as that of Lin et al. (2007), is working toward building assessment tools and scales for measuring spoken language. Geers (2006) and Geers et al. (2011) have also conducted extensive research on the topic, but implantation is still relatively new.

6. Some teams of Deaf researchers are already making headway in such important work that advocates for more inclusive policies to reduce harm to deaf children (e.g., Humphries et al. 2012).

BIBLIOGRAPHY

American Academy of Pediatrics. "Health Issues: Hearing Loss." http://www.healthychildren.org/.

Americans with Disabilities Act of 1990. Pub. L. 101–336. 26 July 1990. 104 Stat. 328.

Anspach, Renée R. 1997. *Deciding Who Lives: Fateful Choices in the Intensive-Care Nursery*. Berkeley: University of California Press.

Anspach, Renée R., and Nissim Mizrachi. 2006. "The Field Worker's Fields: Ethics, Ethnography and Medical Sociology." *Sociology of Health & Illness* 28(6): 713–31.

Apple, Rima D. 1995. "Constructing Mothers: Scientific Motherhood in the Nineteenth and Twentieth Centuries." *Social History of Medicine* 8 (2): 161–78. doi:10.1093/shm/8.2.161.

Archbold, Sue. 2006. "Cochlear Implants and Deaf Education: Conflict or Collaboration?" In *Cochlear Implants*, edited by Susan B. Waltzman and J. Thomas Roland, 183–91. New York: Thieme.

Aronson, Josh. 2001. *Sound and Fury*. Next Wave Films.

Bahan, Benjamin. 2008. "Upon the Formation of a Visual Variety of the Human Race." In *Open Your Eyes: Deaf Studies Talking*, edited by H-Dirksen L. Bauman, 83–99. Minneapolis: University of Minnesota Press.

Baker, Sharon. 2011. *VL2 Integration of Research and Education: Brief 2: Advantages of Early Visual Language*. Washington, D.C.: National Science Foundation Science of Learning Center on Visual Language and Visual Learning.

Barton, Len. 1996. *Disability and Society: Emerging Issues and Insights*. New York: Longman.

Bauchspies, Wenda K., Jennifer Croissant, and Sal P. Restivo. 2006. *Science, Technology, and Society: A Sociological Approach*. New York: Wiley.

Bauman, Dirksen, and Joseph Murray. 2009. "Reframing: From Hearing Loss to Deaf Gain." *Deaf Studies Digital Journal* 1 (1).

Bauman, H.-Dirksen L. 2008. *Open Your Eyes: Deaf Studies Talking*. Minneapolis: University of Minnesota Press.

Baynton, Douglas C. 1998. *Forbidden Signs: American Culture and the Campaign against Sign Language*. Chicago: University of Chicago Press.

Becker, Howard. 1967. "Whose Side Are We On?" *Social Problems* 14: 239–47.

Belzner, Kate A., and Brenda C. Seal. 2009. "Children with Cochlear Implants: A Review of Demographics and Communication Outcomes." *American Annals of the Deaf* 154 (3): 311–33.

Blum, Linda M. 2007. "Mother-Blame in the Prozac Nation Raising Kids with Invisible Disabilities." *Gender & Society* 21 (2): 202–26. doi:10.1177/0891243206298178.

———. 2011. "'Not This Big, Huge, Racial-Type Thing, but . . .': Mothering Children of Color with Invisible Disabilities in the Age of Neuroscience." *Signs* 36 (4): 941–67. doi:10.1086/658503.

Blume, Stuart. 1997. "The Rhetoric and Counter-Rhetoric of a 'Bionic' Technology." *Science, Technology, & Human Values* 22 (1): 31–56.

———. 2010. *The Artificial Ear: Cochlear Implants and the Culture of Deafness*. New Brunswick, N.J: Rutgers University Press.

Boggs, A. 2010. "Spokane Father Won't Force Deaf Daughter to Wear Required Cochlear Implants." *News Tribune* (Spokane, WA), April 29.

Boons Tinne, Jan L. Brokx, Johan M. Frijns, Louis Peeraer, Birgit Philips, Anneke Vermeulen, Jan Wouters, and Astrid van Wieringen. 2012. "Effect of Pediatric Bilateral Cochlear Implantation on Language Development." *Archives of Pediatrics & Adolescent Medicine* 166 (1): 28–34. doi:10.1001/archpediatrics.2011.748.

Bosk, C. 2001. "Irony, Ethnography, and Informed Consent." In *Bioethics in Social Context*, edited by Barry Hoffmaster. Philadelphia: Temple University Press.

Boss, Emily F., John K. Niparko, Darrell J. Gaskin, and Kimberly L. Levinson. 2011. "Socioeconomic Disparities for Hearing-Impaired Children in the United States." *The Laryngoscope* 121 (4): 860–66. doi:10.1002/lary.21460.

Bradham, Tamala S., Geneine Snell, and David Haynes. 2009. "Current Practices in Pediatric Cochlear Implantation." *Perspectives on Hearing and Hearing Disorders in Childhood* 19 (1): 32–42. doi:10.1044/hhdc19.1.32.

Bragg, Lois. 2001. *Deaf World: A Historical Reader and Primary Sourcebook*. New York: NYU Press.

Brown, Kevin D., and Thomas J. Balkany. 2007. "Benefits of Bilateral Cochlear Implantation: A Review." *Current Opinion in Otolaryngology & Head and Neck Surgery* 15 (5): 315–18. doi:10.1097/MOO.0b013e3282ef3d3e.

Brown, Phil. 1995. "Naming and Framing: The Social Construction of Diagnosis and Illness." *Journal of Health and Social Behavior* 35, Extra Issue: 34–52.

Burch, Susan. 2002. *Signs of Resistance: American Deaf Cultural History, 1900 to World War II*. New York: NYU Press.

Burke, Teresa Blankmeyer. 2006. "Comments on 'W(h)ither the Deaf Community?'" *Sign Language Studies* 6 (2): 174–80. doi:10.1353/sls.2006.0015.

Callon, Michel, John Law, and Arie Rip. 1986. *Mapping the Dynamics of Science and Technology: Sociology of Science in the Real World*. Houndmills, Basingstoke, Hampshire: Macmillan.

Campbell, Fiona Kumari. 2009. *Contours of Ableism: The Production of Disability and Abledness*. New York: Palgrave Macmillan.

Casper, Monica J., and Daniel R. Morrison. 2010. "Medical Sociology and Technology Critical Engagements." *Journal of Health and Social Behavior* 51 (1 suppl): S120–32. doi:10.1177/0022146510383493.

Centers for Disease Control and Prevention. 2014. *Preliminary Summary of 2012 National CDC EHDI Data*. http://www.cdc.gov.

Chang, David T., Alvin B. Ko, Gail S. Murray, James E. Arnold, and Cliff A. Megerian. 2010. "Lack of Financial Barriers to Pediatric Cochlear Implantation: Impact of Socioeconomic Status on Access and Outcomes." *Archives of Otolaryngology—Head & Neck Surgery* 136 (7): 648–57. doi:10.1001/archoto.2010.90.

Charmaz, Kathy. 2006. *Constructing Grounded Theory: A Practical Guide through Qualitative Analysis*. New York: SAGE.

Chorost, Michael. 2006. *Rebuilt*. Boston: Houghton Mifflin.

———. 2011. *World Wide Mind: The Coming Integration of Humans and Machines*. New York: Simon and Schuster.

Christiansen, John B., and Irene W. Leigh. 2002. *Cochlear Implants in Children: Ethics and Choices*. Washington, D.C.: Gallaudet University Press.

———. 2010. "Cochlear Implants and Deaf Community Perceptions." In *Cochlear Implants: Evolving Perspectives*, edited by Raylene Paludneviciene and Irene W. Leigh, 39–55. Washington, D.C.: Gallaudet University Press.

Clare, Eli. 1999. *Exile and Pride: Disability, Queerness and Liberation*. Cambridge, Mass.: South End Press.

Clark, Graeme. 2003. *Cochlear Implants: Fundamentals and Applications*. New York: Springer.

Clarke, Adele, Laura Mamo, Jennifer Ruth Fosket, Jennifer R. Fishman, Janet K. Shim, eds. 2010. *Biomedicalization: Technoscience, Health, and Illness in the U.S.* Durham, N.C.: Duke University Press Books.

Clarke, Adele, and Theresa Montini. 1993. "The Many Faces of RU486: Tales of Situated Knowledges and Technological Contestations." *Science, Technology, & Human Values* 18 (1): 42–78.

Clarke, Adele, and Janet Shim. 2011. "Medicalization and Biomedicalization Revisited: Technoscience and Transformations of Health, Illness and American Medicine." In *Handbook of the Sociology of Health, Illness, and Healing*, edited by Bernice A. Pescosolido, Jack K. Martin, Jane D.

McLeod, and Anne Rogers, 173–99. Handbooks of Sociology and Social Research. New York: Springer.
Cochlear Americas Corporation. 2008. *Chief Executive Officer/Chairman Report*. http://www.cochlear.com.
Colligan, Sumi. 2004. "Why the Intersexed Shouldn't Be Fixed: Insights from Queer Theory and Disability Studies." In *Gendering Disability*, edited by Bonnie Smith and Beth Hutchinson. New Brunswick, N.J.: Rutgers University Press.
Conger, Rand D., Katherine J. Conger, and Monica J. Martin. 2010. "Socioeconomic Status, Family Processes, and Individual Development." *Journal of Marriage and the Family* 72 (3): 685–704. doi:10.1111/j.1741-3737.2010.00725.x.
Connolly, William E. 2002. *Neuropolitics: Thinking, Culture, Speed*. Minneapolis: University of Minnesota Press.
Conrad, Peter. 1992. "Medicalization and Social Control." *Annual Review of Sociology* 18 (1): 209–32.
———. 2007. *The Medicalization of Society: On the Transformation of Human Conditions into Treatable Disorders*. Baltimore: Johns Hopkins University Press.
Corker, Mairian, and Sally French. 1999. *Disability Discourse*. Buckingham; Philadelphia: Open University Press.
Crouch, Robert A. 1997. "Letting the Deaf Be Deaf. Reconsidering the Use of Cochlear Implants in Prelingually Deaf Children." *Hastings Center Report* 27 (4): 14–21.
Cyrus, Bainy, Eileen Katz, Celeste Cheyney, and Frances M. Parsons. 2005. *Deaf Women's Lives: Three Self-Portraits*. Washington, D.C.: Gallaudet University Press.
Davey, Monica. 2011. "Among Twists in Budget Woes, Tensions over Teaching the Deaf." *New York Times*, July 26, sec. U.S. http://www.nytimes.com.
Davidson, Kathryn, Diane Lillo-Martin, and Deborah Chen Pichler. 2014. "Spoken English Language Development among Native Signing Children with Cochlear Implants." *Journal of Deaf Studies and Deaf Education* 19 (2): 238–50. doi:10.1093/deafed/ent045.
Davis, Dána-Ain. 2013. "Border Crossings: Intimacy and Feminist Activist Ethnography in the Age of Neoliberalism." In *Feminist Activist Ethnography: Counterpoints to Neoliberalism in North America*, edited by Christa Craven and Dána-Ain Davis, 23–38. New York: Lexington Books.
Davis, Lennard J. 2013. *The Disability Studies Reader*. 4th ed. New York: Routledge.
Desjardin, Jean L. 2005. "Maternal Perceptions of Self-Efficacy and Involvement in the Auditory Development of Young Children with Prelingual

Deafness." *Journal of Early Intervention* 27 (3): 193–209. doi:10.1177/105381510502700306.

Eckert, Richard C. 2010. "Toward a Theory of Deaf Ethnos: Deafnicity ≈ D/deaf (Hómaemon • Homóglosson • Homóthreskon)." *Journal of Deaf Studies and Deaf Education*, enq022. doi:10.1093/deafed/enq022.

Fjord, Lakshmi. 1999. "'Voices Offstage': How Vision Has Become a Symbol to Resist in an Audiology Lab in the U.S." *Visual Anthropology Review* 15 (2): 121–38. doi:10.1525/var.2000.15.2.121.

———. 2001. "Ethos and Embodiment: The Social and Emotional Development of Deaf Children." *Scandinavian Audiology* 30 (2): 110–15. doi:10.1080/010503901750166844.

———. 2010. "Contested Signs: Deaf Children, Indigeneity, and Disablement in Denmark and the United States." In *Deaf and Disability Studies: Interdisciplinary Perspectives*, edited by Susan Burch and Alison Kafer, 67–100. Washington D.C.: Gallaudet University Press.

Flexer, Carol. 2014. "Auditory Brain Development—Language and Literacy Development." Accessed April 23. http://hearinghealthmatters.org.

Fortnum, Heather M., David H. Marshall, and A. Quentin Summerfield. 2002. "Epidemiology of the UK Population of Hearing-Impaired Children, Including Characteristics of Those with and without Cochlear Implants—Audiology, Aetiology, Comorbidity and Affluence: Epidemiología de La Población Infantil de Hipoacúsicos En El Reino Unido, Incluyendo Las Características de Aquellos Con Y Sin Implante Coclear-Audiología, Etiología, Co-Morbilidad Y Nivel Económico." *International Journal of Audiology* 41 (3): 170–79. doi:10.3109/14992020209077181.

Foucault, Michel. 1988. *Technologies of the Self: A Seminar with Michel Foucault*. Amherst: University of Massachusetts Press.

———. 2010. *The Birth of Biopolitics: Lectures at the Collège de France, 1978–1979*. Reprint ed. New York: Picador.

Francis, Ara. 2012. "Stigma in an Era of Medicalisation and Anxious Parenting: How Proximity and Culpability Shape Middle-Class Parents' Experiences of Disgrace." *Sociology of Health & Illness* 34 (6): 927–42. doi:10.1111/j.1467-9566.2011.01445.x.

Frank, Arthur W. 2013. *The Wounded Storyteller: Body, Illness, and Ethics*, 2nd ed. Chicago; London: University of Chicago Press.

Friedner, Michele. 2010. "Biopower, Biosociality, and Community Formation: How Biopower Is Constitutive of the Deaf Community." *Sign Language Studies* 10 (3): 336–47. doi:10.1353/sls.0.0049.

Friedner, Michele, and Stefan Helmreich. 2012. "Sound Studies Meets Deaf Studies." *The Senses and Society* 7 (1): 72–86. doi:10.2752/174589312X13173255802120.

Gallaudet Research Institute. 2003. *Regional and National Summary Report of Data from the 2001–2002 Annual Survey of Deaf and Hard of Hearing Children & Youth (January 2003)*. Washington, D.C.: GRI, Gallaudet University.

———. 2011. *Regional and National Summary Report of Data from the 2009–10 Annual Survey of Deaf and Hard of Hearing Children & Youth (April 2011)*. Washington, D.C.: GRI, Gallaudet University.

Garrett, Peter. 2010. *Attitudes to Language*. New York: Cambridge University Press.

Geers, Ann E. 2006. "Factors Influencing Spoken Language Outcomes in Children Following Early Cochlear Implantation." *Advances in Oto-Rhino-Laryngology* 64: 50–65. doi:10.1159/000094644.

Geers, Ann E., Christine A. Brenner, and Emily A. Tobey. 2011. "Long-Term Outcomes of Cochlear Implantation in Early Childhood: Sample Characteristics and Data Collection Methods." *Ear and Hearing* 32 (1 Suppl): 2S–12S. doi:10.1097/AUD.0b013e3182014c53.

Gentile, Katie. 2011. "What about the Baby? The New Cult of Domesticity and Media Images of Pregnancy." *Studies in Gender & Sexuality* 12 (1): 38–58. doi:10.1080/15240657.2011.536056.

Hall, Kim Q. 2011. *Feminist Disability Studies*. Bloomington: Indiana University Press.

Halpin, Kathy S., Kay Y. Smith, Judith E. Widen, and Mark E. Chertoff. 2010. "Effects of Universal Newborn Hearing Screening on an Early Intervention Program for Children with Hearing Loss, Birth to 3 Yr of Age." *Journal of the American Academy of Audiology* 21 (3): 169–75. doi:10.3766/jaaa.21.3.5.

Haraway, Donna. 1988. "Situated Knowledges: The Science Question in Feminism and the Privilege of Partial Perspective." *Feminist Studies* 14 (3): 575. doi:10.2307/3178066.

Harmon, Amy. 2004. "Neurodiversity Forever; The Disability Movement Turns to Brains." *New York Times*, May 9, Week in Review. http://www.nytimes.com.

Hassanzadeh, S. 2012. "Outcomes of Cochlear Implantation in Deaf Children of Deaf Parents: Comparative Study." *Journal of Laryngology & Otology* 126 (10): 989–94. doi:10.1017/S0022215112001909.

Hauser, Peter C., Amanda O'Hearn, Michael McKee, Anne Steider, and Denise Thew. 2010. "Deaf Epistemology: Deafhood and Deafness." *American Annals of the Deaf* 154 (5): 486–92. doi:10.1353/aad.0.0120.

Haynes, R. Brian, D. Wayne Taylor, and David L. Sackett. 1979. *Compliance in Health Care*. Baltimore: Johns Hopkins University Press.

Herd, Pamela, and Madonna H. Meyer. 2002. "Care Work: Invisible Civic Engagement." *Gender & Society* 16 (5): 665–88. doi:10.1177/089124302236991.

Hintermair, Manfred, and John A. Albertini. 2005. "Ethics, Deafness, and New Medical Technologies." *Journal of Deaf Studies and Deaf Education* 10 (2): 184–92. doi:10.1093/deafed/eni018.

Holden-Pitt, Lisa. 2000. *Gallaudet Research Institute: Who and Where Are Our Children with Cochlear Implants*. Washington D.C.: GRI, Gallaudet University. http://research.gallaudet.edu.

Honigfeld, Lisa, Brenda Balch, and Ann Gionet. 2011. "Early Hearing Detection and Intervention: The Role of the Medical Home." Presented at the 10th Annual Early Hearing Detection & Intervention Conference, Atlanta, February 21.

Hume, Lynne, and Jane Mulcock. 2004. *Anthropologists in the Field: Cases in Participant Observation*. New York: Columbia University Press.

Humphries, Tom, Poorna Kushalnagar, Gaurav Mathur, Donna Jo Napoli, Carol Padden, Christian Rathmann, and Scott R. Smith. 2012. "Language Acquisition for Deaf Children: Reducing the Harms of Zero Tolerance to the Use of Alternative Approaches." *Harm Reduction Journal* 9 (April): 16. doi:10.1186/1477-7517-9-16.

Hyde, Merv, and Des Power. 2006. "Some Ethical Dimensions of Cochlear Implantation for Deaf Children and Their Families." *Journal of Deaf Studies and Deaf Education* 11 (1): 102–11. doi:10.1093/deafed/enj009.

Hyde, Merv, and Renée Punch. 2011. "The Modes of Communication Used by Children with Cochlear Implants and Role of Sign in Their Lives." *American Annals of the Deaf* 155 (5): 535–49.

Hyde, Merv, Renée Punch, and Linda Komesaroff. 2010. "A Comparison of the Anticipated Benefits and Received Outcomes of Pediatric Cochlear Implantation: Parental Perspectives." *American Annals of the Deaf* 155 (3): 322–38.

Ingber, Sara, Michal Al-Yagon, and Esther Dromi. 2010. "Mothers' Involvement in Early Intervention for Children with Hearing Loss." *Journal of Early Intervention* 32 (5): 351–69. doi:10.1177/1053815110387066.

Ingber, Sara, and Esther Dromi. 2010. "Actual versus Desired Family-Centered Practice in Early Intervention for Children with Hearing Loss." *Journal of Deaf Studies and Deaf Education* 15 (1): 59–71. doi:10.1093/deafed/enp025.

Jain, Sarah S. 1999. "The Prosthetic Imagination: Enabling and Disabling the Prosthesis Trope." *Science, Technology & Human Values* 24 (1): 31–54. doi:10.1177/016224399902400103.

Johnson, R. 2006. "Cultural Constructs That Impede Discussions about

Variability in Speech-Based Educational Models for Deaf Children with Cochlear Implants." *Perspectiva* 24: 29–80.

Joint Committee on Infant Hearing. 1994. *Joint Committee on Infant Hearing 1994 Position Statement*. http://www.jcih.org.

Jones, Thomas W., and Karen M. Ewing. 2002. "An Analysis of Teacher Preparation in Deaf Education: Programs Approved by the Council on Education of the Deaf." *American Annals of the Deaf* 147 (5): 71–78. doi:10.1353/aad.2012.0246.

Kafer, Alison. 2013. *Feminist, Queer, Crip*. Bloomington: Indiana University Press.

Kermit, Patrick. 2012. "Enhancement Technology and Outcomes: What Professionals and Researchers Can Learn from Those Skeptical about Cochlear Implants." *Health Care Analysis* 20 (4): 367–84. doi:10.1007/s10728–012–0225–0.

Kirkham, Erin, Chana Sacks, Fuad Baroody, Juned Siddique, Mary Ellen Nevins, Audie Woolley, and Dana Suskind. 2009. "Health Disparities in Pediatric Cochlear Implantation: An Audiologic Perspective." *Ear and Hearing* 30 (5): 515–25. doi:10.1097/AUD.0b013e3181aec5e0.

Knoors, Harry, and Marc Marschark. 2012. "Language Planning for the 21st Century: Revisiting Bilingual Language Policy for Deaf Children." *Journal of Deaf Studies and Deaf Education* 17 (3): 291–305. doi:10.1093/deafed/ens018.

Kuhn, Thomas S. 1996. *The Structure of Scientific Revolutions*. 3rd ed. University of Chicago Press.

Ladd, Paddy. 2003. *Understanding Deaf Culture: In Search of Deafhood*. Clevedon, England ; Buffalo: Multilingual Matters.

Ladd, Paddy, and Harlan Lane. 2013. "Deaf Ethnicity, Deafhood, and Their Relationship." *Sign Language Studies* 13 (4): 565–79. doi:10.1353/sls.2013.0012.

Landsman, Gail. 2008. *Reconstructing Motherhood and Disability in the Age of "Perfect" Babies*. New York: Routledge.

Lane, Harlan. 1989. *When the Mind Hears: A History of the Deaf*. New York: Vintage.

———. 1995. "Constructions of Deafness." *Disability & Society* 10 (2): 171–90. doi:10.1080/09687599550023633.

———. 2005. "Ethnicity, Ethics, and the Deaf-World." *Journal of Deaf Studies and Deaf Education* 10 (3): 291–310.

Lane, Harlan, and Ben Bahan. 1998. "Ethics of Cochlear Implantation in Young Children: A Review and Reply from a Deaf-World Perspective." *Otolaryngology—Head and Neck Surgery: Official Journal of American Academy of Otolaryngology-Head and Neck Surgery* 119 (4): 297–313.

Lane, Harlan, Robert Hoffmeister, and Ben Bahan. 1996. *A Journey into the Deaf-World*. 2nd ed. San Diego: DawnSignPress.

Lareau, Annette. 2003. *Unequal Childhoods: Class, Race, and Family Life*. Berkeley: University of California Press.

Latour, Bruno. 1988. *Science in Action: How to Follow Scientists and Engineers through Society*. Boston: Harvard University Press.

Law, John, and John Hassard. 1999. *Actor Network Theory and After*. Oxford; Malden, Mass.: Wiley-Blackwell.

Law, John, and Annemarie Mol. 2002. *Complexities: Social Studies of Knowledge Practices*. Durham, N.C.: Duke University Press Books.

Leiter, Valerie. 2004. "Dilemmas in Sharing Care: Maternal Provision of Professionally Driven Therapy for Children with Disabilities." *Social Science & Medicine (1982)* 58 (4): 837–49.

Lemke, Thomas. 2011. *Biopolitics: An Advanced Introduction*. New York: New York University Press.

Lewiecki-Wilson, Cynthia, and Jen Cellio. 2011. *Disability and Mothering: Liminal Spaces of Embodied Knowledge*. Syracuse, N.Y.: Syracuse University Press.

Li, Yuelin, Lisa Bain, and Annie G. Steinberg. 2004. "Parental Decision-Making in Considering Cochlear Implant Technology for a Deaf Child." *International Journal of Pediatric Otorhinolaryngology* 68 (8): 1027–38. doi:10.1016/j.ijporl.2004.03.010.

Lin, Frank R., Kristin Ceh, Deborah Bervinchak, Anne Riley, Richard Miech, and John K. Niparko. 2007. "Development of a Communicative Performance Scale for Pediatric Cochlear Implantation." *Ear and Hearing* 28 (5): 703–12. doi:10.1097/AUD.0b013e31812f71f4.

Linton, Simi. 1998. *Claiming Disability: Knowledge and Identity*. New York: NYU Press.

———. 2007. *My Body Politic: A Memoir*. Ann Arbor: University of Michigan Press.

Longmore, Paul K. 2003. *Why I Burned My Book and Other Essays on Disability*. Philadelphia: Temple University Press.

Lovett, R. E. S., P. T. Kitterick, C. E. Hewitt, and A. Q. Summerfield. 2010. "Bilateral or Unilateral Cochlear Implantation for Deaf Children: An Observational Study." *Archives of Disease in Childhood* 95 (2): 107–12. doi:10.1136/adc.2009.160325.

Luterman, David. 1991. *When Your Child Is Deaf: A Guide for Parents*. Parkton, Md.: York Press.

———. 2004. "Children with Hearing Loss: Reflections on the Past 40 Years." *The ASHA Leader*, November 16. http://www.asha.org.

Lutfey, K. E., and W. J. Wishner. 1999. "Beyond 'Compliance' Is 'Adherence.' Improving the Prospect of Diabetes Care." *Diabetes Care* 22 (4): 635–39.

Lyness, C. R., B. Woll, R. Campbell, and V. Cardin. 2013. "How Does Visual Language Affect Crossmodal Plasticity and Cochlear Implant Success?" *Neuroscience & Biobehavioral Reviews* 37 (10): 2621–30. doi:10.1016/j.neubiorev.2013.08.011.

Mairs, Nancy. 1997. *Waist-High in the World: A Life among the Nondisabled*. Boston: Beacon Press.

Mayberry, Rachel I., and Ellen B. Eichen. 1991. "The Long-Lasting Advantage of Learning Sign Language in Childhood: Another Look at the Critical Period for Language Acquisition." *Journal of Memory and Language* 30 (4): 486–512. doi:10.1016/0749–596X(91)90018-F.

Mayberry, Rachel I., and Elizabeth Lock. 2003. "Age Constraints on First versus Second Language Acquisition: Evidence for Linguistic Plasticity and Epigenesis." *Brain and Language* 87 (3): 369–84.

Maynard, Douglas W. 2003. *Bad News, Good News: Conversational Order in Everyday Talk and Clinical Settings*. Chicago: University of Chicago Press.

Meadow, Kathryn P. 1968. "Parental Response to the Medical Ambiguities of Congenital Deafness." *Journal of Health and Social Behavior* 9 (4): 299–309.

Meadow-Orlans, Kathryn P., Patricia Elizabeth Spencer, Lynne Sanford Koester, and Annie G. Steinberg. 2004. "Implications for Intervention with Infants and Families." In *The World of Deaf Infants: A Longitudinal Study*, edited by Kathryn P. Meadow-Orlans, Patricia Elizabeth Spencer, and Lynne Sanford Koester, 218–28. New York: Oxford University Press.

Mertes, Jennifer, and Jill Chinnici. 2011. "Cochlear Implants—Considerations in Programming for the Pediatric Population—The Listening Center at Johns Hopkins Audiology Online." Accessed November 1. http://www.audiologyonline.com.

Metzl, Jonathan M., and Anna Kirkland. 2010. *Against Health: How Health Became the New Morality*. New York: NYU Press.

Mills, Mara. 2012. "Do Signals Have Politics? Inscribing Abilities in Cochlear Implants." In *The Oxford Handbook of Sound Studies*, edited by T. Pinch and K. Bijsterveld, 320–46. Oxford: Oxford University Press.

Milroy, James. 2001. "Language Ideologies and the Consequences of Standardization." *Journal of Sociolinguistics* 5 (4): 530–55. doi:10.1111/1467–9481.00163.

Mitchell, Ross E., and Michael A. Karchmer. 2006. "Demographics of Deaf Education: More Students in More Places." *American Annals of the Deaf* 151 (2): 95–104. doi:10.1353/aad.2006.0029.

Moeller, Mary Pat. 2000. "Early Intervention and Language Development in Children Who Are Deaf and Hard of Hearing." *Pediatrics* 106 (3): e43–e43. doi:10.1542/peds.106.3.e43.

Moeller, Mary Pat, Karl R. White, and Lenore Shisler. 2006. "Primary Care Physicians' Knowledge, Attitudes, and Practices Related to Newborn Hearing Screening." *Pediatrics* 118 (4): 1357–70. doi:10.1542/peds.2006-1008.

Mol, Annemarie. 2002. *The Body Multiple: Ontology in Medical Practice*. Durham, N.C.: Duke University Press.

Moser, Ingunn. 2000. "Against Normalisation: Subverting Norms of Ability and Disability." *Science as Culture* 9 (2): 201–40. doi:10.1080/713695234.

———. 2006. "Disability and the Promises of Technology: Technology, Subjectivity and Embodiment within an Order of the Normal." *Information, Communication & Society* 9 (3): 373–95. doi:10.1080/13691180600751348.

Mundy, Liza. 2002. "A World of Their Own (Deaf Lesbian Couple)." *Waterloo Cedar Falls (IA) Courier.* May 31. http://wcfcourier.com.

Murphy, Catherine. 2009. "Bergen County: A Model Program for Listen and Spoken Language." *Volta Voices* October/November 2009.

National Early Intervention Longitudinal Study. 2007. *National Early Intervention Longitudinal Study: Demographic Characteristics of Children and Families Entering Early Intervention*. New York State Department of Health Division of Family Health Bureau of Early Intervention.

National Institute on Deafness and other Communication Disorders. 2015. "Institute on Deafness and other Communication Disorders—Quick Statistics." http://www.nidcd.nih.gov.

National Institutes of Health. 1995. *NIH Consensus Statement Online: Cochlear Implants in Adults and Children*. 13 (2): 1–30. http://consensus.nih.gov.

New York State Department of Health Division of Family Health Bureau of Early Intervention. 2013. *Part C Annual Performance Report (APR) for FFY 2011 July 1, 2011 through June 30, 2012*. New York State Department of Health Division of Family Health Bureau of Early Intervention.

Nikolopoulos, Thomas P., Hazel Lloyd, Sue Archbold, and Gerard M. O'Donoghue. 2001. "Pediatric Cochlear Implantation: The Parents' Perspective." *Archives of Otolaryngology—Head & Neck Surgery* 127 (4): 363–67.

Niparko, John K., Emily A. Tobey, Donna J. Thal, Laurie S. Eisenberg, Nae-Yuh Wang, Alexandra L. Quittner, Nancy E. Fink, for the CDaCI Investigative Team. 2010. "Spoken Language Development in Children following Cochlear Implantation." *JAMA: The Journal of the American Medical Association* 303 (15): 1498–1506. doi:10.1001/jama.2010.451.

Okubo, Suguru, Miyako Takahashi, and Ichiro Kai. 2008. "How Japanese Parents of Deaf Children Arrive at Decisions regarding Pediatric

Cochlear Implantation Surgery: A Qualitative Study." *Social Science & Medicine (1982)* 66 (12): 2436–47. doi:10.1016/j.socscimed.2008.02.013.

Oliva, Gina A. 2004. *Alone in the Mainstream: A Deaf Woman Remembers Public School*. Washington, D.C.: Gallaudet University Press.

Oliver, Michael. 1990. *The Politics of Disablement: A Sociological Approach*. London: Palgrave Macmillan.

———. 2013. "The Social Model of Disability: Thirty Years On." *Disability & Society* 28 (7): 1024–26. doi:10.1080/09687599.2013.818773.

Padden, Carol A., and Tom L. Humphries. 1990. *Deaf in America: Voices from a Culture*. Cambridge, Mass.: Harvard University Press.

Parens, Erik. 2000. *Enhancing Human Traits: Ethical and Social Implications*. Washington, D.C.: Georgetown University Press.

———. 2008. *Surgically Shaping Children: Technology, Ethics, and the Pursuit of Normality*. Baltimore: Johns Hopkins University Press.

———. 2011. "On Good and Bad Forms of Medicalization." *Bioethics* 27 (1): 28–35. doi:10.1111/j.1467–8519.2011.01885.x.

Parens, Erik, and Adrienne Asch, eds. 2000. *Prenatal Testing and Disability Rights*. Washington, D.C.: Georgetown University Press.

Peterson, Nathaniel R., David B. Pisoni, and Richard T. Miyamoto. 2010. "Cochlear Implants and Spoken Language Processing Abilities: Review and Assessment of the Literature." *Restorative Neurology and Neuroscience* 28 (2): 237–50.

Petitto, Laura-Ann. 2014. "Advances in Human Language Acquisition—Insights from the Petitto Brain and Language Laboratory for Neuroimaging (BL2) and Visual Language and Visual Learning (VL2)." Washington D.C., May 1. http://webcast.gallaudet.edu.

Pickersgill, Martyn. 2011. *Sociological Reflections on the Neurosciences*. Bingley, U.K.: Emerald Group Publishing Limited.

Pickersgill, Martyn, Sarah Cunningham-Burley, and Paul Martin. 2011. "Constituting Neurologic Subjects: Neuroscience, Subjectivity and the Mundane Significance of the Brain." *Subjectivity* 4 (3): 346–65. doi:10.1057/sub.2011.10.

Pitts-Taylor, Victoria 2010. "The Plastic Brain: Neoliberalism and the Neuronal Self." *Health: An Interdisciplinary Journal for the Social Study of Health, Illness and Medicine* 14 (6): 635–52. doi:10.1177/1363459309360796.

Preisler, Gunilla, Anna-Lena Tvingstedt, and Margareta Ahlstrom. 2005. "Interviews with Deaf Children about Their Experiences Using Cochlear Implants." *American Annals of the Deaf* 150 (3): 260–67. doi:10.1353/aad.2005.0034.

Rapp, Rayna. 2000. *Testing Women, Testing the Fetus: The Social Impact of Amniocentesis in America*. New York: Routledge.

———. 2010. "Chasing Science: Children's Brains, Scientific Inquiries, and Family Labors." *Science, Technology & Human Values* 36 (5): 662–84. doi:10.1177/0162243910392796.

Rapp, Rayna, and Faye Ginsburg. 2001. "Enabling Disability: Rewriting Kinship, Reimagining Citizenship." *Public Culture* 13 (3): 533–56.

Reagan, T. 2002. "Toward an 'Archeology of Deafness': Etic and Emic Constructions of Identity in Conflict." *Journal of Language, Identity & Education* 1 (1): 41–66. doi:10.1207/S15327701JLIE0101_4.

Riessman, Catherine K. 1983. "Women and Medicalization: A New Perspective." *Social Policy* 14 (1): 3–18.

Rose, Nikolas. 2006. *The Politics of Life Itself: Biomedicine, Power, and Subjectivity in the Twenty-First Century*. Princeton, N.J.: Princeton University Press.

Rose, Nikolas, and Joelle M. Abi-Rached. 2013. *Neuro: The New Brain Sciences and the Management of the Mind*. Princeton, N.J.: Princeton University Press.

Rothman, Barbara Katz. 1993. *The Tentative Pregnancy: How Amniocentesis Changes the Experience of Motherhood*. New York: W. W. Norton and Company.

Rousso, Harilyn. 2013. *Don't Call Me Inspirational: A Disabled Feminist Talks Back*. Philadelphia: Temple University Press.

Russ, Shirley A., Doris Hanna, Janet DesGeorges, and Irene Forsman. 2010. "Improving Follow-Up to Newborn Hearing Screening: A Learning-Collaborative Experience." *Pediatrics* 126 (Supplement): S59–69. doi:10.1542/peds.2010–0354K.

Sass-Lehrer, Marilyn, and Barbara Bodner-Johnson. 2003. "Early Intervention: Current Approaches to Family-Centered Programming." In *Oxford Handbook of Deaf Studies, Language, and Education*, edited by Marc Marschark and Patricia Elizabeth Spencer. Oxford: Oxford University Press.

Savarese, Emily Thornton, and Ralph James Savarese. 2009. " 'The Superior Half of Speaking': An Introduction." *Disability Studies Quarterly* 30 (1). http://dsq-sds.org.

Saxton, Marsha. 2000. "Why Members of the Disability Community Oppose Prenatal Diagnosis and Selective Abortion." In *Prenatal Testing and Disability Rights*, edited by Erik Parens and Adrienne Asch. Washington D.C.: Georgetown University Press.

Schur, Edwin M. 1972. *Labeling Deviant Behavior; Its Social Implications*. New York: Joanna Cotler Books.

Scott, Pam, Evelleen Richards, and Brian Martin. 1990. "Captives of Controversy: The Myth of the Neutral Social Researcher in Contemporary Scientific Controversies." *Science, Technology & Human Values* 15 (4): 474–94. doi:10.1177/016224399001500406.

Scully, Jackie Leach. 2008. *Disability Bioethics: Moral Bodies, Moral Difference*. London: Rowman and Littlefield.

Shakespeare, Tom. 2006. *Disability Rights and Wrongs*. New York: Routledge.

Shakespeare, Tom, and Nicholas Watson. 2001. "The Social Model of Disability: An Outdated Ideology?" *Research in Social Science and Disability* 2 (June): 9–28. doi:10.1016/S1479–3547(01)80018-X.

Shapiro, Joseph P. 1994. *No Pity: People with Disabilities Forging a New Civil Rights Movement*. New York: Broadway Books.

Siebers, Tobin Anthony. 2008. *Disability Theory*. Ann Arbor: University of Michigan Press.

Sinclair, Jim. 1993. "Don't Mourn for Us: A Letter to Parents." *Neurodiversity*. neurodiversity.com. Accessed December 20, 2014.

Sismondo, S. 2008. "Science and Technology Studies and an Engaged Program." In *The Handbook of Science and Technology Studies,* edited by Edward J. Hackett, 13–32. 3rd. ed. Cambridge, Mass.: MIT Press; published in cooperation with the Society for the Social Studies of Science.

Snoddon, Kristin. 2012. *American Sign Language and Early Literacy: A Model Parent-Child Program*. Washington, D.C.: Gallaudet University Press.

Snodden, Kristin, and Kathryn Underwood. 2014. "Toward a Social Relational Model of Deaf Childhood." *Disability & Society* 29 (4): 530–42. doi:10.1080/09687599.2013.823081.

Solomon, Andrew. 1994. "Defiantly Deaf." *New York Times*, August 28, Magazine. http://www.nytimes.com.

Sorkin, Donna L. 2013. "Cochlear Implantation in the World's Largest Medical Device Market: Utilization and Awareness of Cochlear Implants in the United States." *Cochlear Implants International* 14 (Suppl 1): S4–12. doi:10.1179/1467010013Z.00000000076.

Sparrow, Robert. 2005. "Defending Deaf Culture: The Case of Cochlear Implants." *Journal of Political Philosophy* 13 (2): 135–52. doi:10.1111/j.1467–9760.2005.00217.x.

Stern, Ryan E., Bevan Yueh, Charlotte Lewis, Susan Norton, and Kathleen C. Y. Sie. 2005. "Recent Epidemiology of Pediatric Cochlear Implantation in the United States: Disparity among Children of Different Ethnicity and Socioeconomic Status." *The Laryngoscope* 115 (1): 125–31. doi:10.1097/01.mlg.0000150698.61624.3c.

Strauss, Anselm, and Juliet M. Corbin. 1990. *Basics of Qualitative Research: Grounded Theory Procedures and Techniques*. 2nd ed. Newbury Park, Calif.: Sage Publications.

"Strivright Auditory Oral School of NY." 2014. Oraldeafed.org.

Tait, M., T. P. Nikolopoulos, L. De Raeve, S. Johnson, G. Datta, E. Karltorp, E. Ostlund, et al. 2010. "Bilateral versus Unilateral Cochlear Implantation in Young Children." *International Journal of Pediatric Otorhinolaryngology* 74 (2): 206–11. doi:10.1016/j.ijporl.2009.11.015.

Timmermans, Stefan, and Marc Berg. 2003. "The Practice of Medical Technology." *Sociology of Health & Illness* 25 (3): 97–114. doi:10.1111/1467-9566.00342.

Timmermans, Stefan, and Mara Buchbinder. 2012. *Saving Babies? The Consequences of Newborn Genetic Screening*. Chicago: University of Chicago Press.

Tucker, Bonnie Poitras. 1998. "Deaf Culture, Cochlear Implants, and Elective Disability." *Hastings Center Report* 28 (4): 6–14. doi:10.2307/3528607.

Valente, Joseph Michael, Benjamin Bahan, and H.-Dirksen L. Bauman. 2011. "Sensory Politics and the Cochlear Implant Debates." In *Cochlear Implants: Evolving Perspectives*, edited by Raylene Paludneviciene and Irene W. Leigh, 245–58. Washington D.C.: Gallaudet University Press.

Van Cleave, Jeanne, Karen A. Kuhlthau, Sheila Bloom, Paul W. Newacheck, Alixandra A. Nozzolillo, Charles J. Homer, and James M. Perrin. 2012. "Interventions to Improve Screening and Follow-Up in Primary Care: A Systematic Review of the Evidence." *Academic Pediatrics* 12 (4): 269–82. doi:10.1016/j.acap.2012.02.004.

Van Cleve, John V. 1989. *A Place of Their Own: Creating the Deaf Community in America*. Washington D.C.: Gallaudet University Press.

Vidal, Fernando. 2009. "Brainhood, Anthropological Figure of Modernity." *History of the Human Sciences* 22 (1): 5–36. doi:10.1177/0952695108099133.

Vrecko, Scott. 2010. "Neuroscience, Power and Culture: An Introduction." *History of the Human Sciences* 23 (1): 1–10. doi:10.1177/0952695109354395.

Wheeler, Alexandra, Sue Archbold, Susan Gregory, and Amy Skipp. 2007. "Cochlear Implants: The Young People's Perspective." *Journal of Deaf Studies and Deaf Education*, January. doi:10.1093/deafed/enm018.

"WHO | International Classification of Functioning, Disability and Health (ICF)." 2014. *World Health Organization*. Accessed July 2. http://www.who.int.

Wilson, Elizabeth A. 2004. *Psychosomatic: Feminism and the Neurological Body*. Durham, N.C.: Duke University Press Books.

Winner, Langdon. 1980. "Do Artifacts Have Politics?" *Daedalus* 109 (1): 121–36.

Wood, Caitlin, ed. 2014. *Criptiques*. San Bernardino, Calif.: May Day Publishing.

Woodcock, Kathryn. 2001. "Cochlear Implants vs. Deaf Culture?" In *The Deaf World*, edited by Lois Bragg, 325–32. New York: New York University Press.

Wrigley, Owen. 1997. *The Politics of Deafness*. Gallaudet University Press.

Yoshinaga-Itano, Christine. 2003a. "Early Intervention after Universal

Neonatal Hearing Screening: Impact on Outcomes." *Mental Retardation and Developmental Disabilities Research Reviews* 9 (4): 252–66. doi:10.1002/mrdd.10088.

———. 2003b. "From Screening to Early Identification and Intervention: Discovering Predictors to Successful Outcomes for Children with Significant Hearing Loss." *Journal of Deaf Studies and Deaf Education* 8 (1): 11–30. doi:10.1093/deafed/8.1.11.

Zola, Irving K. 1972. "Medicine as an Institution of Social Control." *Sociological Review* 20 (4): 487–504.

INDEX

LAURA MAULDIN is assistant professor of human development and family studies, and women's, gender, and sexuality studies, at the University of Connecticut. She is a nationally certified American Sign Language interpreter.